ISLANDS
Portraits of Miniature Worlds

Louise B. Young

W. H. Freeman and Company
New York

Illustrations: Jennifer Dewey
Design: Victoria Tomaselli

Cataloging-in-Publication data on file.

Printed in the United States of America
First Printing: April, 1999

To my grandchildren
in the hope that their generation
will be more successful than ours
in balancing the desires of mankind
with the needs of all the other forms
of life that share this beautiful
planet with us

Contents

Foreword

I have always loved islands. Without exactly analyzing why, I have chosen to visit many islands around the world and have felt a deep pleasure in experiencing the unique properties of each one. It was not until I began to think about writing a book on this subject that I became aware of the characteristics that are common to all of them: the isolation, the physical beauty, the fascinating endemic species. At the same time, I recognized that every one of my favorite islands was significant in its own right. Each one gives us a glimpse of a simpler world, as the Earth was before it became so densely inhabited by mankind.

My choice of islands for this book was based on these perceptions. I have not included islands that have played a political role in human history, like England or Japan or Manhattan. They have been so transformed by the presence of mankind and so modernized that they no longer exemplify island habitats. Instead, I have chosen islands that still retain a distinctive personality or whose story illuminates a significant aspect of man's relationship with nature. And I have visited almost all of these islands myself.

Thirty-five years ago, after I survived a life-threatening illness, my husband and I decided that we would no longer put off the far-flung travel that we had always hoped to do—someday. Every year thereafter we have taken a trip to some different, exotic part of the world. This travel we almost always did "on our own," researching the special significance of our destination, the means of getting there, and the highlights that we could experience in each of these faraway places. Once we arrived, we usually rented a car and drove ourselves through the country, stopping to talk to the people and to spend time in any place we found especially interesting. When I speak of "we" in this book, I mean the two of us. We rarely joined any group, except in the few cases where individual travel was unusually dangerous or even impossible.

Many vivid memories crowd my memory. I have piloted a glider plane over the beautiful coastline of Oahu; from a single-engine plane I have seen the incredible shades of blue that delineate the bottom of the Bahama Sea and looked down into the fiery Parícutin volcano in Mexico during the earliest days of its eruption. I have watched a Komodo dragon tear great pieces of flesh from a freshly killed goat; and I have held a gentle ring-tailed lemur in my arms. I have watched a pride of 30 lions walk in solemn single file past my canvas tent; and I have fished for piranhas on a lake in the rain forests of Peru. I have watched the rising sun gild the snowcapped peaks of the Annapurna mountain range and have seen the green flash as the sun set beneath the Aegean Sea.

For these and all the other wonders I am deeply grateful. On mountains and on deserts, on land and on sea, I have seen an incredible variety of glorious landscapes and unusual communities of living things. But the special places—the ones that meant the most to me—were islands. I want to share these impressions with others who love islands as much as I do.

These stories touch on a number of scientific fields. In order to verify the relevant facts and theories, I needed the help of many experts. I am especially indebted to the following scientists for their assistance:

Alfred M. Ziegler and Fred Anderson, professors of geology, and David Rawley, associate professor of geology—all at the University of Chicago; Paul Sereno, professor of organistic biology and anthropology at the University of Chicago; and Timothy Earle, professor of anthropology at Northwestern University. I am grateful for their generous gift of time and expertise. But the final responsibility, of course, still lies with me.

Louise B. Young
April, 1999

Introduction

If once you have slept on an island
You'll never be quite the same
You may look as you looked the day before
And go by the same old name.

You may bustle about the street or shop,
You may sit at home and ponder.
But you'll see blue water and wheeling gulls
Wherever you may wander.

You may chat with neighbors of this and that
And close to your fire keep,
But you'll hear ship's whistle and lighthouse bell
And tides beat through your sleep.

Oh you won't know why and you can't say how
Such change upon you came;
But once you have slept on an island
You will never be quite the same.

Throughout human history, islands have held a special place in the minds, the imaginations, the affections of mankind. What are the special secrets of their charm? The remoteness of islands surrounds them with a certain mystery, and their isolation is responsible for their individual characteristics and evolutionary history.

Many islands are endowed with great natural beauty on a scale that is accessible to human beings. Within an hour or a day, one can climb soaring mountains, walk on beaches of coral sand, and experience the many moods of the all-encircling sea. No view of the ocean from an island is exactly like any other. The ocean provides infinite variety, from slow, gentle movement to a wildly disturbed surface with steep waves curling in green arcs and exploding with a roar in a chaotic mass of foam.

The embracing presence of the ocean is one of the most important factors in the island environment. It moderates the island climate, warming the air in winter and cooling it in summer. Ocean water has several properties that are very advantageous for islands. It changes temperature very slowly, and more heat is needed to raise the temperature of a given quantity of water than is necessary for any other known liquid. Similarly, as heat is dissipated, the temperature of water drops very slowly. This gradual warming and cooling characteristic produces a remarkably stable environment, and it is responsible for an interesting phenomenon that occurs every 24 hours in the atmosphere above an island. As the sun shines on the land and on the sea, it warms the surface of the island more rapidly than the water around it. This causes air to rise above the island, and when this air mass reaches the colder upper atmosphere, the moisture in it condenses into clouds. Thus, most islands in the afternoon are marked by the presence of a cloud lying just downwind of the land. This phenomenon has been observed by navigators for centuries. The Polynesian adventurers used it as a tool to find their way from island to island over the vast stretches of the Pacific Ocean. The existence of the islands of New Zealand was guessed at many years before they were discovered and were known as the land of the "long white cloud."

Scientists have recognized recently that the remarkable attributes of water made it a very favorable medium for the earliest living things—and, in fact, that it was the cradle of life. Although these conclusions were reached by scientists, there is within each of us an instinctive knowledge of the primal significance of water. Many people hold water sacred. They are baptized in it; they pour it over their heads at sunrise; they touch it to their foreheads in prayer.

On the smaller islands there is no place where the presence of water cannot be seen or felt or heard. The regular pounding of waves on the shore is deeply reassuring, like the beating heart an infant feels in his mother's arms.

On a sunny day on a warm tropical island, the dominant sensation is the color blue—a tone that suggests both beauty and mystery. The waters are blue with an intensity that defies their crystal clarity. Looking straight down into their depths, one can see vivid moving pictures of the colorful and varied life within these waters. Anyone who has observed the underwater life in these clear waters has been fascinated by the enormous variety— an everchanging kaleidoscope of color and shape and form. And yet, looking out across the surface of the sea, the presence of this underworld is hidden from view. The only visual impression is many shades of blue, from pale aqua to deep cobalt.

The sky, too, is blue on a fine day, and although it seems transparent stretching from the Earth far off into space, it also hides another world—a universe of stars and galaxies, of pulsars and black holes. Blue is an ethereal color, guarding great mysteries in worlds beyond our own.

⬥

The landscape of islands varies greatly, depending on the circumstances of their birth from the sea. Many have erupted by volcanic action suddenly and violently, bringing fresh matter from deep inside the Earth to the planet's surface. New earth, new water, and new air are created in these fiery furnaces. This earth is rich in many essential ingredients, such as lime, potash, and phosphates, and as it weathers, it makes a fertile soil in which many types of vegetation flourish. On Sicily, Kauai, and Sri Lanka, for example, one can witness the wealth of luxuriant growth and spectacular crops produced by this rich earth.

The manner in which nature produced these new bits of the Earth's crust has been illuminated by discoveries on the ocean bottom. The sudden appearance of islands has actually been observed and their history recorded.

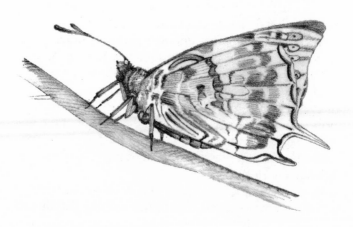

One of the classic examples was an island that erupted from the sea southwest of Sicily and then disappeared again a few months later. On June 28, 1831, the crew of a British ship in that area reported that they felt a sudden jolt, as though the ship had run aground. But subsequent observations showed that they had experienced an earthquake at sea. About two weeks later, the captain of another vessel, en route to Agrigento in Sicily, saw a tremendous column of steam, like an enormous waterspout. On July 18, returning from Agrigento, the ship passed by the same spot and found that an island had risen out of the sea. It was about 12 feet above water level with a volcanic crater in the center ejecting great plumes of vapor and projectiles of molten rock. Inside the circular crater a pool of murky red liquid boiled and seethed.

By August 4 the island was 300 feet high and several miles in circumference. After that day, the volcanic activity of the main vent subsided, but violent agitation of the sea southwest of the island occurred, producing columns of dense white steam. Geologists concluded that a second vent had erupted but had not ejected enough solid matter to reach the surface. In the meantime, the waves, the winds, and the rain beating on the small new island were rapidly eroding its surface and carrying it back to the sea. By the end of October, there was only a little mound of volcanic rock left above water, and two years later it was entirely submerged.

In recent years many drowned islands like this have been observed below the surface of the ocean. Hundreds of truncated islands have been identified on the seafloor. Their tops range from 3000 to 6000 feet below present sea level. Most of them have a flattened summit, as though the peak of a cone had been sliced off. These flat-topped seamounts (or guyots) appear to be old volcanic islands. As the volcanic activity abated, the matter that had been forced up receded. And when sea level had reached within a few hundred yards of the peak, wave action began to erode the land that was still emergent. Gradually, the peak was planed off. Continued subsidence and erosion drowned the islands completely.

It is an interesting observation that almost all volcanoes are either in the sea or no farther than 150 miles from its edge. This fact can be understood in terms of the geological theory of plate tectonics. The lithosphere (which contains the crust and the solid part of the Earth's mantle) appears to be divided into 12 or so large pieces that have been moving in relation to each other very slowly throughout time. In general, volcanic action is concentrated along the boundaries of these lithospheric plates, often at midocean ridges where the plates are drifting apart and seafloor spreading is taking place. It also occurs in the regions where plates are colliding and some crustal material is being forced down again into the mantle. As this material descends to great depths, it is heated and becomes partially melted. Wherever this semiliquid magma finds a crack or weak spot in the crust, it moves upward again. The lithosphere beneath the oceans is much younger and thinner than the continental crust, which is very old, and, because of gradual changes in the chemistry of the rocks, has grown slowly thicker with time. It is not surprising, therefore, that volcanoes are most often born in or near the sea.

Volcanic islands usually create a surrounding area in the sea that is shallow enough to allow the penetration of sunlight through the water. Where that water is warm, it provides a favorable environment for the formation of coral communities, and these conditions exist in a circle around the island. As time passes, the volcanic land becomes more and more compacted and eroded, causing the island to subside. Finally, the island sinks entirely beneath the water, leaving a ring-shaped coral atoll cupping a little segment of the sea. The land that now exists above water is no longer volcanic in character. It is biological and quite different in chemistry from the volcanic soil.

A number of islands, however, are quite dissimilar from the volcanic ones. Much older, they were originally part of an ancient continental shelf. All the large land masses on the Earth are surrounded by shelves that vary greatly in width, from less than a half mile off the west coast of Peru to more than 800 miles wide off Siberia. Taken all together, the continental shelves occupy one-tenth of the world's surface. They have a granitic base like the continents (not the basaltic crust that is characteristic of the ocean floor) that is overlaid by great accumulations of soil washed down from higher land.

At certain periods in the Earth's history, when glaciers locked up large amounts of the planet's water supply, the oceans were lower and the continental shelves were dry land. Then, as the Earth warmed, the sea levels rose slowly day by day, century by century. The lowest areas were inundated by the sea and the highest areas became islands. The Bahamas are classic examples of islands formed in this way, while the Virgin Islands are almost all volcanic in origin.

A few islands, of which Madagascar is the most prominent example, were created in still another way— by the rupture of a large land mass. Tectonic activity resulted in the separation of a piece of land from its parent continental crust, and the severed piece, carrying its cargo of living things, drifted farther and farther from the mainland. The species that lived on the new island evolved in their own distinctive ways. Depending on the time when the separation occurred and the species that happened to be present at that time, evolution produced forms of life quite unlike those that evolved on the parent land mass, perhaps unlike any others that exist on Earth.

The scale of an island's ecology is a very significant factor in creating its special image. A little island develops an individuality, a personality of its own. On these small pieces of land surface, a few species have been isolated, perhaps by tectonic activity or perhaps coming together by chance, brought by the winds, the tides, and the ocean currents. On each island a unique community has evolved, and over the centuries the interaction of the species with the environment has generated new life forms.

In this way, isolation has spawned exotic life forms. Some of them are beautiful, like the bird of paradise and the angel's-trumpet. Others are dangerously destructive, like the Komodo dragon or the bird-eating tree snake of

the Solomon Islands. Isolation has also been responsible for the unfavorable results of genetic inbreeding, producing an island where color blindness is very common in the human population and an island where harelips are frequently seen. These phenomena are symptomatic of inbreeding and are caused by a doubling of recessive alleles that may undermine the general health of a species. This condition has been called "inbreeding depression" and is believed to be one of the factors leading to the extinction of a species.

In fact, biologists have surmised that the phenomenon of extinction is directly dependent on the number of individuals of each species present in a community. Fifty is estimated to be the minimum size for survival—the number to avoid inbreeding. A much larger number of individuals is required to maintain long-term adaptability, the successful response to a changing environment. Five hundred has been suggested as the number of individuals that would provide sufficient genetic variety. The 50/500 rule has caught the attention of biologists and conservationists around the world, and it is part of a movement to make ecology a more exact and mathematical science. New insights have occurred as a result of this movement. For example, the number of species an island can support depends on the size of the island. Large islands like Madagascar harbor many more species than little ones, such as Aldabra in the Seychelles. In fact, the chances that endemic species will become extinct are greater on a small island.

These generalizations, of course, do not take into consideration certain important differences within populations, such as the nature of each individual species, the balance between the sexes, the number of nubile females, and the breeding habits. But they are useful in planning for conservation projects and methods of protecting endangered species. They emphasize the point that extinction begins to threaten long before the population of breeding individuals approaches zero and that the success of a species depends on the presence of a community of like individuals.

But while island communities may spawn forms of life that have a poor chance of survival, they are also the agents for change and variety, adding enormously to the interest of natural history. The study of life would be a comparatively dull affair without such exotic forms of life as the dodo bird, the platypus, the aardvark, and the ring-tailed lemur.

The fascinating flora and fauna that can be found on islands have added wonder to the island image. The excitement of travel and exploration has been enhanced by the discovery of the unusual and bizarre. Without this inducement the journeys of world travelers like Charles Darwin and Alfred Wallace might not have occurred at all and certainly would not have yielded the insights that have increased mankind's understanding of nature and its own place in it. A world averaged out so that every part of it is like every other would be a much less interesting and challenging place.

Moreover, significant facts about nature and ourselves can be learned by examining these miniature worlds and their histories. The simplified ecology of an island reveals principles of evolution that would be much more difficult to understand by studying the vastly more complex history of life on the continents. (While continents harbor millions of species, some islands harbor as few as 30.)

Although most islands share the qualities of isolation, beauty, wonder, and the magic of the surrounding sea, each one is an individual with its own story to tell.

Some of the stories are tragic, others are hopeful, and all of them convey important messages about the delicate balance of living communities and the consequences of the ways we exploit or nurture our own little island Earth.

Chapter 1

THE HAWAIIAN ARCHIPELAGO

The Garden Isles

A green isle in the sea, love,

A fountain and a shrine,

All wreathed with fairy fruits and flowers,

And all the flowers were mine.

—EDGAR ALLAN POE

The Hawaiian Islands are the very quintessence of "islandness." They are isolated, they harbor many exotic and endemic species, and they possess great beauty colored with mystery and wonder.

The Hawaiian archipelago is a long chain of volcanic islands wending north northwest across the Pacific. At the southern end of the chain is the island of Hawaii, which is actively volcanic today. Mauna Loa, which has erupted on average every 3.7 years throughout the twentieth century, is one of the most active volcanoes on Earth. On Maui, the next island to the northwest, the volcanoes are extinct and slightly eroded. Then come Molokai and Oahu, with volcanoes that are long extinct and deeply eroded. Kauai, Nihoa, Necker, Lisianski, Midway, and Kure Atoll lie in a sequence to the northwest. Beyond Kure stretches a chain of drowned islands—guyots and other seamounts that rise steeply from a submarine ridge.

The ages of the islands have been determined, and they fall in this same order: Hawaii is the youngest, at half a million years; Oahu is 4 million years old; Kauai, about 5 million; Midway, 18 million; and Kure Atoll is 20 million years of age. This regular succession suggests that the Hawaiian chain formed as the plate carrying the Pacific Ocean passed in a northwesterly direction over a plume of concentrated heat in the Earth's mantle. Twenty million years ago the plume was beneath Kure Atoll, while half a million years ago the plume was beneath Hawaii. Today, a new island, Loihi, is being formed seventeen miles southeast of Hawaii.

Using modern technology, scientists are able to observe and monitor the birth throes of this new island. They have descended in a submersible to a half mile below the ocean surface, where they have watched the changing topography of the wildly disturbed summit of Loihi. Craggy recesses, toppled rock formations, and large volcanic boulders strewn over 4 or 5 miles testify to a riot of landslides. Geologists are concerned that major eruptions at Loihi might cause a tsunami, a huge shock wave, that could devastate the coast of Hawaii. There is danger even in the submersible, because currents caused by the volcanic activity tend to push the vessel up on one side of Loihi's crater and suck it down on the other.

Loihi is now more than 2.8 miles above the ocean floor, a growth that has taken millions of years, and Loihi is not expected to reach the surface for another ten thousand years or so. In the meantime, the crater continues to belch forth superheated water, carbon dioxide, minerals, and clouds of tiny microbes, primitive organisms that thrive in this extreme environment.

The sequence of the Hawaiian activity is not just an interesting phenomenon. It has been so compelling a fact that the hypothesis of a hot spot in the Earth's mantle has been generally accepted by geologists, even though it seems to contradict the theory that volcanoes are caused only by lithospheric plates colliding or separating. (The Hawaiian Islands are close to the center of the Pacific plate, so there is no possibility that they were created due to movement at the edge.) Moreover, other examples of plumes have been found around the world. Perhaps their existence will lead to a deeper understanding of the powerful but still mysterious forces that are moving pieces of the Earth's crust around the planet's surface.

While volcanism is an ever-present reality on the island of Hawaii, it proceeds slowly enough to allow human beings to move to safe ground, but it is still an awesome process. Mauna Loa and Kilauea frequently spurt forth fountains of fire. From lakes of glowing lava, long ribbons of incandescent matter creep slowly down the flanks of the mountains, threatening and sometimes engulfing little settlements. When the lava congeals, it is dark and wrinkled like an elephant's trunk and ugly, as are most newborn things.

The immensity and significance of this geologic activity dwarfs our human scale. Early inhabitants of Hawaii believed it must be the action of a god. Pelé was the name given to this female deity by the Polynesians. She was portrayed as tall and slender with long blonde hair and fair skin and as a capricious and jealous god to be propitiated with gifts precious to the Hawaiian people: fine silk scarves, *Ohi a-lehua* blossoms, berries of the *ohelo* tree, and libations of *awa*, an intoxicating drink made from the roots of a pepper tree. Despite these gifts, eruptions continued without warning.

When these new bits of the Earth's crust emerged suddenly from the sea, they were far removed from other land masses. For many months, these black rocks were as stark and barren as the moon, devoid of all life. Then, as blazing sunshine alternated with drenching rains, the harsh surfaces slowly began to soften. Winds brought a varied assortment of life forms. Spiders sailed in on strands of silken webs; these tiny voyagers are frequently seen miles above the Earth's surface in such great numbers that at times the air takes on an iridescent sheen. These spiders were among the first forms of life to take up residence on islands newly born out of the sea.

Spores light enough to float on the breezes were carried thousands of miles from more ancient lands and deposited at random across the bare mountain flanks. A few of these spores found a toehold on the dark, forbidding rocks and grew and began to work their transformation upon the land. Lichens were probably the first successful flora. These are not single individual plants; each one is a symbiotic combination of alga and fungus. The algae capture the sun's energy by photosynthesis and store it in organic molecules. The fungi absorb moisture and mineral salts from the rocks, passing these on in waste products that nourish the algae.

It is significant that the earliest living things that built communities on these islands are examples of symbiosis, a phenomenon that depends upon the close cooperation of two or more forms of life. As we will see, this principle carries over throughout the entire process.

Lichens helped to speed the decomposition of the hard rock surfaces, preparing a soft bed of soil that was abundantly supplied with minerals that had been carried in the molten magma from the bowels of the Earth. Now, other forms of life could take hold—ferns and mosses that flourish even in rocky crevices.

These plants propagate by producing spores, tiny fertilized cells that contain all the instructions for making a new plant, but the spores are unprotected by any outer coating and carry no supply of nutrient. Vast numbers of them are cast loose to scatter on the ground beneath the mother plants. Sometimes they are carried a little farther afield by water or by wind. But only those few spores that settle down in very favorable locations can start new life; the vast majority fall on barren ground. By force of sheer numbers, however, the mosses and ferns survived and multiplied. Some species developed great size, becoming tree ferns that even now grow in the Hawaiian forests.

It was many millions of years later that another kind of flora evolved on Earth—the seed-bearing plants. This was a wonderful biological invention. The seed has an outer coating that surrounds the genetic material of the new plant, and inside this covering is a concentrated supply of nutrients. Thus, the seed's chances of survival are greatly enhanced over that of the naked spore.

Long before the Hawaiian Islands were born from the sea, an even more favorable type of seed-bearing plant had appeared. Known as *angiosperms*, these plants include all forms of blooming vegetation. In the angiosperm the seeds are wrapped in an additional layer of covering. Some of these coats are hard, like the shell of a nut, for extra protection. Some are soft and tempting, like a peach or a cherry. In some angiosperms the seeds are equipped with gossamer wings, like the dandelion and milkweed seeds. These new characteristics offered better ways for the seeds to move to new habitats. They could travel through the air, float in water, and lie dormant for many months.

Plants with large buoyant seeds—like coconuts—drift on the ocean currents and are washed up on the shores. Remarkably resistant to the vicissitudes of ocean travel, they can survive prolonged immersion in salt water. When they come to rest on warm beaches, and the conditions are favorable, the seed coats soften. Nourished by their imported supply of nutrients, the young plants push out their roots and establish their place in the sun.

By means of these seeds, plants spread more widely to new locations, even to isolated islands like the Hawaiian archipelago, which lies more than 2000 miles west of California and 3500 miles east of Japan. Seeds of grasses, flowers, and blooming trees made the long trips to these islands. (Grasses are simple forms of angiosperms that bear their encapsulated seeds on long stalks. They are the ancestors of such basic cereal crops as wheat and rice.)

In a surprisingly short time, angiosperms added a rich palette of colors to the vegetation and filled many of the land areas that had been bare. Sweet, nutritious foods hung in tempting profusion from the trees—bananas, papayas, and breadfruit, to name just a few. Nuts lay strewn on the ground, and grains carried a compacted form of energy. Flowers found the new islands an ideal habitat. Nurtured in the subtropical warmth and brilliant sunshine, they grew to enormous size. A paradise of lush beauty took shape where only a little while ago dark lava had erupted from a boiling sea.

Birds and insects blew in on the wind, too—not all varieties, of course, just those that happened to catch the right breeze and survive the long journey across the sea. The Hawaiian Islands are in the belt of the trade winds, which are gentle and predictable surface winds caused by differential heating by the sun of various portions of the Earth's surface and by the rotation of the planet. Temperature differences cause the surface air to flow toward the equator from about 30 degrees in both hemispheres, and, at the same time, the Earth is turning westward, creating an easterly component to the winds. The trade winds are predominantly northeasterly in the northern hemisphere and southeasterly in the southern. Except during storms, these winds are steady and consistent over the oceans, blowing at about 11 to 13 miles an hour. The Hawaiian Islands, at 20 to 25 degrees north latitude, are directly in the path of the trade winds, which provide very pleasant weather conditions and ideal transportation for various forms of life.

Because these islands are so isolated, the plants and animals that became established there evolved in unique ways. The communities of living things that formed on the Hawaiian Islands were different from the communities that evolved on continents. In the early years of the islands' formation, the number of species was much smaller and the competition between them was much less severe. There was abundant food, water, sunlight, and space available for all. There were no large mammals, no carnivorous beasts. In fact, the conditions were especially favorable for the survival of many new forms of life.

Like a talented composer, nature is constantly experimenting with new variations on old established themes. Mutations in the genetic code that determines the shape and characteristics of each species occur frequently throughout time. They are caused by radiation, by chemicals, by cosmic rays, or by other factors that have not yet been identified. The vast majority of mutations are so deleterious that the resulting organisms do not survive, and marginal ones are weeded out by natural selection—the survival of the fittest—in a highly competitive environment. But under the favorable conditions of the Hawaiian Islands in their early years, a few of the new variations were able to survive and develop in their own individual ways. Thus, species were produced that were endemic.

Wingless and blind insects evolved in these islands. Some birds, like the rails, found such an abundant ground-based food supply that they had no need to travel far, and they lost the power of flight. Land snails became tree snails. Many shrubs and flowers, finding little competition from larger vegetation, grew tall and produced unusual varieties of blooming trees, like lobelias, hibiscus, and tree violets. In other parts of the world, gardeners know lobelia as a small, colorful ornamental. In the Hawaiian Islands, under less competitive stress, the native lobelias have radiated into many tall and unusual forms. There are 120 different types endemic to Hawaii.

On the other hand, several familiar types of life were conspicuously missing. No truly freshwater fish evolved in the Hawaiian Islands, nor did any native amphibians, reptiles, or mammals, except for one species of bat. Large insects were not successful in colonizing the islands, but flowers were abundant, and special types of birds evolved to take over the role of pollination. From one original progenitor there are now 23 species of honey creepers, dainty birds with long, curved bills to sip the nectar from the angel's-trumpet, the *ohe naupaka*, and the cup of gold.

Here again we witness the evolution of symbiotic relationships. New species were created that served the needs of other forms of life and in so doing opened up an important opportunity for themselves. A unique and closely interrelated colony of living things grew and flourished.

—

And while all this was taking place on land, in the sea surrounding the islands another finely integrated community was taking shape—the coral reef. Like the lichen, the basic structure of the coral reef is built upon a remarkable relationship between two quite different forms of life. The tiny coral polyp is an animal. It must obtain its energy by the ingestion of organic molecules in which the sun's energy has been stored by plants using the process of photosynthesis. Capturing zooplankton from the sea is one way of obtaining this nourishment, but coral has another much more reliable food supply. It grows its own vegetable garden—millions of single-celled algae called *zooxanthellae*. And the coral reef in its turn provides the algae with food and shelter. They feed upon the waste produced by the metabolism of the coral polyps. The algae capture the energy of the sun, synthesize organic compounds, and thus produce an ideal source of food for the coral. This relationship, beneficial to both partners, is so successful that coral reefs have multiplied and grown wherever conditions are favorable for their formation. The water in which they grow must be shallow enough so that the zooxanthellae receive a considerable amount of sunshine. A fast-growing reef is usually no deeper than 150 feet. And coral cannot thrive in cold waters. The temperature must be at least 75 degrees Fahrenheit.

The coral reef provides shelter and safe hiding places for a multitude of living organisms. Soft-bodied gorgonians find the reef an ideal platform on which to erect their graceful fans where the sun shines and the rough pounding of unprotected sea is moderated. Echinoderms, sponges, sea worms, and anemones take up residence in this sheltered environment. The reef also provides food and protection for many species of fish. They feed on the algae and other organisms that capture and store the energy of the sun. Their waste products contain phosphates and nitrates that are consumed and excreted in turn by the coral polyps. Then they are passed to the zooxanthellae, providing these important nutrients to the algae.

＿

As time went on, however, the introduction of foreign species into the Hawaiian Islands changed these relationships both on land and on sea. The advent of man changed the composition of the communities that had evolved by increasing the competitive pressures enormously and thereby endangering the most fragile of the endemic species.

The first human settlers in Hawaii were Polynesians who came by canoe, probably from the Society Islands around A.D. 500. They brought along sweet potatoes and taro, which they cultivated for poi, their favorite staple food. They also introduced dogs and chickens, the dingo, and a Polynesian species of small pig. With these invasions some aspects of the environment changed forever. Humans and animals killed off several species of endemic birds, including the world's only flightless ibis. The colonists not only introduced alien plant species but also destroyed much of the lowland forest to make room for agriculture.

Nevertheless, these ancient Hawaiians managed to live for more than a thousand years in relative harmony with the land, even though their population increased rapidly. When the islands were visited by Captain James Cook in 1778, the human population had reached several hundred thousand. To accommodate that number with minimal environmental damage, the natives must have established a harmonious relationship with the land and the other species that shared it.

In the cultural heritage of the early Hawaiians, the use of resources was regulated through a strict taboo system and religious beliefs. The catching of certain fish, for example, was suspended at various times during the year to honor the gods with whom those fish were associated. The practical result was that the fish population had a chance to replenish itself. Many forms of nature were believed to be attributes of the major gods. Each Hawaiian had a personal or family relationship with a certain god and took special care not to harm any of the "body forms" of that god. If you were of the Pelé clan, for example, you were related to one of Pelé's brothers, a shark god, and the shark would be sacred to you.

The upland forests were considered to be holy places, the "forests of the gods." Hawaiians trapped birds in them, gathered plants, and occasionally cut down a large tree to make a temple idol or a canoe, but they did not clear cut in order to cultivate the land. The ecosystem of the rain forests was not seriously disturbed.

Although the early Polynesian settlers did in general treat their environment with respect and love, they practiced several customs that took a toll on certain native species. For example, the bright plumage of beautiful endemic birds was harvested and used as decoration, jewelry, symbols of prestige, and even currency. One chieftain possessed a spectacular yellow cape composed of the feathers of 80,000 mamo, a bird with brilliant yellow tail feathers.

When the Europeans came to the Hawaiian Islands, beginning in 1778 with the landing of Captain Cook, many other alien species were introduced: goats and sheep, cattle and pigs, the mongoose, the domestic cat, and the axis deer from India. Rats that had hooked a ride on the ship came ashore and stayed. These species multiplied, and some, like the deer, browsed so destructively that professional hunters were hired to cull their number.

Some alien species that were deliberately introduced to relieve an environmental problem created a greater problem than the one they had been expected to resolve. The casuarina tree was imported from Australia and planted in many areas for watershed protection—to bind the soil in order to prevent erosion and sometimes to provide an effective windbreak. However, it grew into thick stands that choked out other vegetation, and the needles poisoned the soil, making it an unsuitable environment for new growth. Guava trees, which bear tasty fruit and lovely white flowers, were introduced in the early 1800s and are another troublesome pest. They are so prolific that they crowd out most other plants.

Although many imported species are undesirable, several have contributed to the success of the community. The swamp mahogany, which like the casuarina was introduced for watershed protection, is one of the largest and most useful trees. It grows rapidly and produces excellent lumber suitable for floors, stakes, and many other building purposes.

The kukui (candlenut) tree was imported by early Polynesians. It has become one of the most important trees in the Hawaiian forests. It grows to 80 feet or more and has light green leaves and small white flowers that highlight the mountain slopes of all the major islands. The bark and the acrid juice of the nut have been used as a dye for tattoos, canoe paint, tapa cloth, and fish nets. In addition, the nuts are strung together, burned as a light source, and pressed for oil. The wood is used for making canoes and fish net floaters. The kukui is now the state tree of Hawaii.

A number of native and endemic species are still present, contributing to the variety and beauty of the landscape. Several of these are survivors of very ancient life forms, for example, the tree ferns. The small variety grows on new lava flows as well as in rocky gulches and on wetter mountain slopes. The distinctive shapes of the fern leaves add interesting patterns to the Hawaiian scene. The soft, fine hair that grows on the erect fronds of new growth has long been used as pillow and mattress stuffing.

The native plants also have provided important medications to treat human illnesses and alleviate pain for centuries. The potential benefits of these plants have barely been sampled. Important wonder drugs may still lie concealed within the shining dark leaves and brilliant petals of the native vegetation.

In interesting and subtle ways, many species contribute to the health and well-being of others who share the same environment. For example, the presence of flowers has a spiritual impact on the nature of man. They delight the eye and lighten the heart, and their beauty is an antidote to stress and anger. Their presence smooths away the tension that develops in response to conflict or fear.* To look into the heart of an opening flower is like listening to great music. The saying in the islands is *"Olu pua"* (feeling peaceful with flowers).

---

*It is an interesting fact that during the terrible period known as the Cultural Revolution in China, the raising of flowers commercially was prohibited. The cultivation of flowers was considered effete, and many private gardens were destroyed as part of the "reeducation" of the people. I visited China just a year after the revolution came to an end, and I noticed the sparsity of flowers. I cannot remember ever seeing a butterfly. I finally understood when I heard that the raising of flowers had been strictly curtailed for ten years—a fact our guides had not mentioned.

The benefits that flowers have provided for the human spirit have been returned by mankind. Flowers—both endemic and imported—have been planted where they have space to grow, and they have been regularly provided with food and water, love and nurturing. This might be considered a special kind of symbiotic relationship.

. The Hawaiian Islands are extraordinarily rich in flowers. The brilliant colors, exotic shapes, and fragrances of many flowers, as well as hundreds of blooming trees, delight the senses. Examples are plumeria, orchids, night-blooming cereus, birds-of-paradise, amaryllis, water lilies, coral hibiscus, passion flowers, and blue torch ginger.

The presence of modern man, however, has disturbed the delicately balanced communities that existed before his arrival. Unlike the early Polynesian settlers, whose relationship with nature was determined by religious beliefs and taboos, the people who came from Western industrial societies were motivated by different goals. Economic advantage had become the greatest good, and all other considerations became subordinate. The changes wrought by these human activities proceeded slowly at first, with small, hardly noticeable increments from one year to the next. But, continued over centuries, they have altered the organization of the Hawaiian Islands ecology offshore as well as on the islands themselves.

On the eastern side of Oahu, at the foot of the Koolau Range, Kaneohe Bay receives the drainage from ten rivers that cascade down the precipitous mountainsides. Land has been cleared along this coast for housing and roads, making it more susceptible to very rapid erosion. When heavy rains fall, the swollen rivers carry a thick load of red volcanic soil into the ocean. The island seems to be bleeding into the sea. For many years, sewage outfalls from this development fouled the waters of the bay, which was once the site of spectacular reefs.

Known as the Coral Gardens, this formation was one of the special treasures of the island. Unfortunately, the presence of man has destroyed this treasure.

Coral reefs have a natural mechanism for cleaning off the sediment that falls on their surfaces. A thin coating of mucus secreted by the coral polyps traps the particles, which are then moved down and off the reef by millions of waving cilia. But this process cannot deal effectively with the thick layer of sediment that occurs when human sewage and silt runoff foul the waters. The cilia are unable to move the heavier load; the zooxanthellae, deprived of light, die; and the reef loses its principal source of nourishment.

Sewage damages the dynamic integrity of the living reef in another way. Nutrients contained in the effluents encourage the growth of an alien type of algae that multiplies prodigiously and can no longer be kept in check by the natural defenses of the reef community. In Kaneohe Bay, green bubble algae took over like a spreading cancer, engulfing entire coral heads. Continuous sheets of this warty growth enveloped broad areas of what was once a magnificent piece of living lithosphere. The smothered coral polyps died, and even the limestone began to dissolve beneath the algae. Recognition of these problems led to corrective measures to restore the health of the reefs. Sewage was diverted to the open ocean, and grading ordinances have reduced erosion.

Despite these corrective measures, humans are continuing to make serious inroads on the Hawaiian reefs in other ways. Aggressive fishing methods are responsible for great areas of dead reef. Strong chemicals are sprayed into the coral caves and recesses where aquatic organisms make their homes, and they are driven out into nets in vast numbers. Reefs that have been sprayed never recover.

It is difficult to judge the impact of all these factors on the ecology of the Hawaiian Islands. We must consider each island as a whole; its health can be recognized by its vitality and beauty.

I had the opportunity once to see the island of Oahu from the perspective of space. On a diamond-bright Hawaiian day, I piloted a glider plane from a small airport on the northern shore of the island. When my glider was released from its tow, I was carried aloft by a steady breeze. The trade wind that sweeps in over a thousand miles of ocean meets the Waianae Range and turns upward. I rode this gentle, consistent wind, banking and turning as the eagles fly.

Beneath me the deep blue sea was etched by long parallel lines of surf-like bars of music, white against blue. In a little bay where the surface was quieter, the water was amber-colored, the characteristic signature of coral reefs below. I turned and flew toward the green mountain slopes, higher and higher, suspended in a shining bubble of air, cradled by a great quietness. The voice of the wind was just a gentle whisper as it flowed softly over the wings.

The hillsides that had looked like green velvet from a distance were now detailed in many shades of green and a profusion of shape and texture. There were dark groves of swamp mahogany, the spreading canopies of monkeypod and banyan trees festooned with curtains of koali vine, and a soft understory of Hawaiian tree ferns beneath coconut palms whose fronds trembled and glittered like a thousand jewels as they turned back the sunshine. Deep in the hollow of each ravine, rows of kukui trees made pale ribbons like glacial streams flowing down the deep lush valleys to the sea.

As I followed the ridge westward the whole leeward coast came into view. Inland I saw mile upon mile of pineapple plantations with young plants set in neat rows against red soil and vast seas of silvery green sugarcane fields ripe for harvest. Soon I could see the far coastlines of the island: the south shore with the three deeply indented lobes of Pearl Harbor; and the misty east shore, veiled in clouds hanging low upon the hulking shapes of the Koolau Range, where erosion is still making a small spreading stain in the sea.

Seeing the whole island thus, complete and self-contained, I recognized that there is still great beauty here. The fact of its continued vitality suggests that the damage has not passed the point of no return. With cooperation, the resilience of nature could restore the equilibrium of its finely tuned ecology. But the timing is crucial. If development proceeds at a faster pace than our ability to control erosion and pollution, the fertile topsoil will end up in the sea, the reefs will be smothered, and the flowers will wither and die. The delicate balance that has sustained the magic of symbiosis will be threatened if mankind continues to disrupt the remarkable communities of living things on both land and sea that have created one of the loveliest places on Earth.

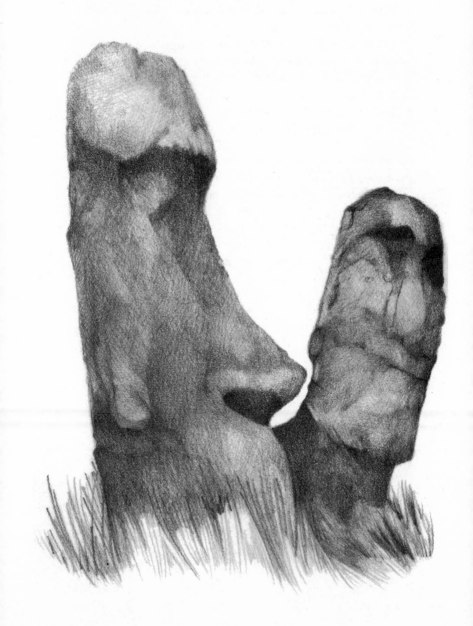

Chapter 2

EASTER ISLAND

A Paradigm for Man

Lord, what fools these mortals be!

—WILLIAM SHAKESPEARE

$\mathcal{E}$aster Island is a little mote of land set in the middle of the widest marine solitude, the emptiest extent of ocean there is in the world. This extreme isolation has been at least partly responsible for the island's tragic history. Most of the delightful attributes of island environments were destroyed there by centuries of human conflict from which there was no escape. The typical island beauty with lush vegetation and many endemic species was erased, leaving a forbidding coastline and desolate bare brown hills. Now, mystery hangs over the island like the fog that envelops it throughout most of the year.

Easter Island carries the ruins of a unique culture that flourished there four or five hundred years ago. During the classic period of this civilization, a strange and dramatic sight was created. Along the shoreline, crowning the steep wave-lashed cliffs, great stone platforms were erected and on these ceremonial altars stood giant statues carved from brown volcanic tuff. Many were capped with red stone topknots. Several hundred of these effigies dominate the scant 35 miles of shoreline. They stood 35 to 60 feet tall, each one weighing 50 tons or more. Facing inward with their backs to the sea, the austere faces with pursed lips and hypnotic eyes made of white coral and black obsidian look out with stern and melancholy gazes at this tiny piece of the Earth's surface, a triangular island pockmarked with the craters of great volcanoes.

The statues seem to be guarding the land—perhaps from the storms that breed in the austral seas and gather momentum as they travel across many thousands of miles, or perhaps from the awesome forces that lie hidden deep within the Earth itself and erupt in sudden violence, spreading fire and incandescent lava across the land.

But no historic record exists to document the visual impact of this scene. The first European ship stopped at Easter Island in 1722, after the old culture had been destroyed and the time of troubles had begun.

The origin and meaning of the great stone effigies, which are called *moai* in the native language, is still a mystery. The skill used in carving them and the building of the ceremonial platforms (*ahus*) suggest a higher cultural heritage than has been evidenced in any other island of the South Pacific. It would seem that the people responsible for this work also had some knowledge of astronomy. Several of the ahus are so oriented that their facades face the rising or setting sun at the vernal and autumnal equinoxes or the summer or winter solstice.

A form of writing was known, too, because hieroglyphics were carved on pieces of wood ("talking sticks" or *rongo-rongo* tablets). No other form of ancient writing has been found in all of Polynesia, and while much effort has been spent in attempts to decipher this writing, all attempts failed until recently. In 1995, Steven Fischer, an anthropologist from New Zealand, succeeded in identifying the significance of a frequently repeated sequence of three hieroglyphics that he translated as "X copulated with Y and the issue was Z." In other words, it was a record of the genealogy of the people.

These are a few of the facts that suggest an origin for the Easter Island people that is different from the peoples who settled the islands farther west. There are some similarities with Incan culture, as the Norwegian explorer and anthropologist Thor Heyerdahl pointed out. For example, the precisely fitted stonework of the ahus is reminiscent of examples of ancient stone walls in Cusco, Peru. Several forms of vegetation, such as the bottle gourd and sweet potato, are typical South American plants. Totora reeds, like those that grow on Lake Titicaca on the Peru-Bolivia boundary and are used to build reed boats, are also found on Easter Island. On the other hand, the language and the physical characteristics of the natives are Polynesian.

Scientists have recently unearthed facts that only deepen the mystery. Several hundred of the many skeletons found on the island have been measured, X-rayed, and classified by anthropologists. They found distinct variations in the shapes of kneecaps and pelvic girdles. These findings were interpreted as suggesting that at least two separate groups settled on Easter Island. However, this assumption has been challenged by other anthropologists, and the question remains unresolved.

Another very interesting fact was discovered by scientists who examined blood samples of living Easter Islanders. An unusual characteristic showed up in these samples—a characteristic that is found in the blood of only one other population group in the world, the Basque people, whose homeland is the northwest corner of the Spanish peninsula. It is hard to imagine how there could be any connection between the Basques and Easter Islanders half the world away, but one possibility has been suggested. In the early sixteenth century, Spanish galleons sometimes carried crews of Basque sailors, and several of these ships were lost, perhaps in the waters near Easter Island. Could some of these shipwrecked men have found their way to the island? One of the traditional wooden carvings found on the island is the figure of a man so emaciated that all his ribs protrude. Perhaps this "statue with ribs" (or *moai kavakava* in the local tongue) represents starving shipwrecked sailors who reached these lonely shores.

Science now has the tools to provide a definitive answer to these questions. DNA taken from samples of ancient human bones on the island may reveal the genetic origin of these peoples. Results of the first studies of this kind indicate a Polynesian ancestry, but this finding does not rule out the possibility that two different racial types coexisted on this little island.

While the origin of this fascinating primitive culture is still speculative, the events that led to its destruction are now quite well understood. It is a story of important human interest, because it shows what can happen to a promising culture when population growth puts great stress on limited natural resources and fighting breaks out over the remaining supplies of food, fuel, and water. The horrifying disintegration of the social structure of these proud and successful people depicts in microcosm the essence of the dilemma of modern man. Isolated from all other land masses, the inhabitants of Easter Island were entirely dependent on its own internal resources.

Easter Island is volcanic in origin. Like Hawaii, it erupted in a cloud of smoke and fire from the bottom of the sea, sometime in the past 50,000 years, and subsequent eruptions of the volcano Rano Raraku continued to occur and altered the shape of the land. Geologists believe that the most recent eruption may have occurred within the last 1000 to 2000 years. It is from the igneous rock at the heart of this volcanic cone that the great stone figures were carved.

The eruptions that created Easter Island were very violent in nature; the lava emerged with great explosive force, ejecting showers of "volcanic bombs," which now lie in profusion across the rolling countryside. These bombs are made of porous black rock, about the size of coconuts. They obstruct travel and are a hindrance to agricultural development.

In three of Easter Island's volcanic cones, crater lakes have formed. They contain freshwater and were once used as a source of drinking water. Now the lakes are stagnant, with their surfaces largely covered with green algae and totora reeds.

It is not surprising that this island bears so many marks of its volcanic origin. It lies next to one of the most active rift zones in the world—the East Pacific Rise. Here the Earth's crust is moving at a faster rate than anywhere else on the planet. The two plates are separating from each other, and new crust is being formed in the rift between them. At the same time, the plate on which Easter Island rides is being subducted beneath the west coast of South America. Since 1988, NASA has been monitoring the change in position of Easter Island by means of laser beams reflected off a satellite. They find that the island is moving toward South America at an average rate of about 3 inches a year.

Easter Island is at a latitude of 27 degrees south and is bathed by the cool Humboldt Current. The waters here are not warm enough to provide ideal conditions for coral growth, and there is no fringing reef like the ones that surround the shores of many South Pacific islands. The sea pounds in against steep cliffs. There is no sheltered lagoon and no protected anchorage. The island, however, enjoys a favorable climate. The average temperatures are in the ideal range between the high 50s and the mid-70s, varying only a few degrees between winter and summer. Moreover, the rainfall is abundant, averaging 56 inches a year—almost twice as much as that which falls on the rich corn belt of the United States.

Given the physical characteristics of the island, one would expect to find a pleasant landscape with varied vegetation—many trees, pasture lands, perhaps fields of sugarcane, and blooming plants—so it is surprising to find a desolate landscape covered with brown, parched grasses. No streams, lakes, or forests break the stark outline of the brown hills against the sky. It is a scene that bears the scars of the impact of man.

There is abundant evidence that in its original state Easter Island was heavily forested. Cores taken from the volcanic craters have revealed the presence of pollen from as many as 20 species of trees, including large deciduous varieties, palms, and conifers. Many of these trees were useful to man, such as mahogany and the indigenous toromiro, which were both used for building purposes and for sculpture. The bark of the paper mulberry was used to make tapa cloth. Another tree, perhaps the coconut palm, yielded fiber for ropes and fishing nets. The nuts of the sandalwood tree were a staple of the early Easter Island diet, and the tree's light-colored wood was much prized for its sweet fragrance.

But early inhabitants were so prodigal in the use of these resources that the island was gradually stripped of its forest growth at a time when the population was large and increasing. These factors are closely related not only on Easter Island but everywhere in the world. As more people must be supported, more land must be cleared to grow crops and more wood is needed as fuel to cook the food. The forests are cut down, erosion occurs, and the land becomes less able to yield abundant crops.

The soil of Easter Island, being of volcanic origin, originally contained important nutrients—lime, potash, phosphates. But this soil lies in a thin layer over very porous lava formations that are like a giant pumice stone. The abundant rain, falling on land without forest cover, runs swiftly through the thin layer of soil, leaching out some of the essential ingredients, and flows down through the porous rock formation to the sea.

Forests provide important protection from erosion and help to build a thicker layer of soil. Tree roots strike

deep and are strong enough to crack the rock formations, gradually turning them into soil. The leaves of the
trees break the impact of every shower. They hold the
water in drops and little puddles, allowing it to drip softly
down to the forest floor, so the soil can absorb it slowly.
Trees shade the soil from the direct rays of sunlight,
reducing the rate of evaporation, and the land remains
moist for a long period of time. Then the water can collect in little streams that flow down beyond the forests,
carrying the benefits of freshwater to less wooded lands.
The leaves that fall from the trees carpet the forest floor
with a protective shield against wind and weather. As
they decay, they become a soft friable mulch that slowly
builds up into more soil and restores important nutrients. Finally, the trees break the impact of strong winds
that destroy tender young growth and further dessicate
the land.

When Easter Island became stripped of forest
growth, it began to suffer from chronic drought and was
raked by the winds that sweep unimpeded over thousands of miles of open sea. Today, less than an hour after
every heavy shower, the land is dry, but the ocean all
around the island is stained dark brown with the soil
that has been carried with the rain out to sea. This influx
of freshwater contaminated with eroded soil is a very
unfavorable environment for fish and drives them away
from shore.

The destruction of the forests on Easter Island, of
course, did not take place all at once. It must have
occurred gradually over several centuries. And this period
of time seems to have coincided with the period when
most of the great moai were carved, between the fourteenth and the seventeenth centuries. At that time, the
population must have been large to provide the manpower for carving and moving these enormous stone
effigies. There are at least 500 of them on the island, and

it has been estimated that a crew of 20 men would take almost a year to create each one. The manpower required to move one of these statues is estimated to have taken another crew of about the same size many months. The people of Easter Island did not have the wheel and they had no metal tools. For carving, they used chisels made of basalt rock and obsidian, a volcanic glass that can be flaked into razor-sharp cutting edges. Although quite rare around the world, obsidian is found in abundance on Easter Island, and the principal source is the volcano Mount Orito on the southwestern part of the island.

All of this work must have taken place in a highly disciplined and organized society. Because the island's food resources were not great even in times when it enjoyed a stand of forest growth, the work of planting, cultivating, and harvesting crops to feed a large population must have been carefully planned and strictly carried out. It seems likely that a repressive society had evolved, with a small group of managers and a large laboring class.

These assumptions are supported by the stories that have been passed down from generation to generation of Easter Islanders. There were, according to these tales, two distinct classes of people: the Long-ears, or Hanau Eepe, who hung heavy ornaments in their earlobes stretching them almost to the shoulders; and the Short-ears, or Hanau Momoko, who did not engage in this practice. The Hanau Momoko were much more numerous than the Hanau Eepe, who represented a small privileged autocracy. The system, although dictatorial, worked well enough to provide sustenance for many people and countless hours of labor for carving and transporting the great stone moai. But as the population continued to grow, more trees were cut down to clear more land for agriculture. The moving of the moai required the use of long wooden poles to lever the heavy statues and to slide them over the rough terrain. Wood was also needed to build boats for fishing and as fuel to cook the food. More trees were harvested and the land became less fertile, more susceptible to periods of drought, and less able to yield good crops.

Around 1678, according to oral history, the Hanau Eepe ordered the Hanau Momoko to pick up all the volcanic bombs that lay strewn over the island and throw them into the sea, thus clearing more land for agriculture. This new demand on their labor was the spark that fired the rebellion of the Hanua Momoko. All work suddenly ceased on the moai. Hundreds of them were left unfinished at the quarry and along the slopes of Rano Raraku. Stone tools were cast aside and still lie scattered there.

Then a terrible battle was fought on the hillside known as Poike Point. Excavations and carbon dating have verified the oral history of this brutal conflict. The

Hanau Eepe assembled on the promontory, which was surrounded on three sides by water, and, at the land side, at the base of the hill, they dug a series of trenches to protect themselves from invasion—making a primitive Maginot Line. They filled the trenches with brush and logs, thus further stripping the forests. Their plan was to conceal themselves behind the piles of dirt thrown up by the excavations and to set the brush on fire in the event of an attack. But the plan backfired.

A Hanau Eepe man who lived on Poike Point had a Hanau Momoko woman working for him as a cook. She was loyal to her own people and sent them a signal when all the Eepe men were asleep. She led two bands of Momoko warriors along the shore around both ends of the line of trenches. A considerable force remained in front of the line, and at dawn they launched a frontal attack. The brush was set on fire. But the Momoko warriors who had penetrated behind the lines attacked from the rear and drove the Eepe men toward the flaming trenches. Continuing their drive, they forced the retreating Eepe warriors into the fires, "as if they were stones," relates the oral history as reported by Father Sebastian Englert. "The trenches were filled and the good odor of the cooked meat of the Hanau Eepe rose into the air." To this day the ditch that marks the site is called *Te Omu O Te Hanau Eepe* (the earth oven of the Hanua Eepe). Only one Eepe man survived this holocaust. Allowed to live, he married a Momoko woman and had many descendants.

Recent excavations have found charred remains in the ditch. Carbon dating of the remains has verified a date of 1678 plus or minus a hundred years. Local tradition sets the date at about 1680. This decisive and brutal battle did not purge the hatred and desire for revenge from the souls of the Momoko. It led to even more terrible conflicts. When the repressive regime of the Eepe was wiped out, the whole fabric of the society was destroyed and anarchy prevailed. The victorious Momoko broke up into tribal groups that turned on each other, fomenting almost continuous warfare. No one was safe away from his kin group. Families hid in the many caves that honeycombed the island or barricaded themselves in stone houses with doors so small that to enter the houses one had to crawl on all fours.

Labor in the fields and gardens was dangerous and was considered degrading, because all organized labor was associated with the time of the autocratic rule of the Eepe. The food supply dwindled until it was insufficient to feed the considerable population that had survived. Raids on neighboring tribes for food were a constant source of conflict. Cannibalism, which may have been practiced earlier but only to a limited extent, now became widespread. Gnawed human bones found heaped in the caves of Easter Island bear witness to the grand scale of this practice. One of these caves is still known as "the cave where men are eaten."

For almost two centuries the conflict continued, becoming ever more violent as the food supply dwindled. In the battles that took place between kin groups, the victors burned the small grass and reed houses of the vanquished and carried the people away as slaves— men, women, and children. When their period of health and usefulness was finished, they were subjected to barbaric tortures. They were beaten with clubs or slashed with obsidian blades. Some were burned over slow fires or trampled until their intestines were ground into the dirt.

During this time all the objects that had been held sacred were desecrated. One by one the great statues were toppled onto their faces and the ceremonial ahus were destroyed. A new religious cult was introduced— probably by another wave of immigrants from Polynesia. Once a year an elaborate ceremony took place when the sooty terns began to nest on the tiny islets just off the coast of Easter Island. Young men representing each of the tribal groups, raced each other to the place where the terns were nesting, and the one to carry back in his mouth the first-laid egg handed it to the warrior chief of his tribe, who was then named the Birdman of the Year. He and his kin group had power over all the other kin groups on the island—a power symbolized by a heavy wooden club more than 3 feet in length and elaborately carved. All the rest of the population concealed themselves in caves to escape slavery and torture.

The final tragedy occurred about 1860. During a three-year period, Peruvian ships came repeatedly to Easter Island to capture and carry off slaves. They seized all the leaders of the community—the strongest men as well as those who could read and write the hieroglyphics on the rongo-rongo tablets. These were the guardians of whatever knowledge remained concerning the origin and evolution of this culture. It is estimated that somewhere between 900 and 1000 slaves were taken and forced to work at digging guano for fertilizer on the Chincha Islands just off the coast of Peru and 2000 miles away. The great majority of them died of overwork and diseases for which they had no immunity. In response to protests lodged with the government of Peru by family members on Easter Island, the remaining 100 survivors were shipped back, but during the voyage many died. Only 15 were alive when the ship arrived at Easter Island, and these survivors carried the germs of smallpox and tuberculosis. Epidemics swept through the remaining population, reducing it to only a few hundred people. This was all that remained of a population of approximately 10,000 that are believed to have lived on the island during the height of that culture.

It was during these centuries of internal conflict that the first contacts with Europeans were made. Half a dozen European expeditions stopped at Easter Island during the eighteenth century, and, although most of them stayed only a few hours, they all reported a barren and desolate landscape. In their accounts there is evidence of a steadily deteriorating condition.

In 1722 a Dutch fleet captained by Jakob Roggeveen visited the island on Easter Day and named it in honor of that occasion. An account written by Carl Friedrich Behrens, who was a member of the expedition, reported that the outward appearance of the island was not inviting. Covered with gray "parched-up grass and charred brushwood . . . it suggested no other idea than that of a sparse and meager vegetation." But the Dutch were greeted by a large crowd of natives, "brightly clad." A great many canoes came to the ship carrying men who snatched the hats and the caps off the heads of the sailors. A party of crewmen that went ashore found several enormous stone heads standing near the shore with their backs to the sea. They were disappointed to find no sign of running streams or ponds offering promise of freshwater to replenish the ship's supply. When asked for food, the natives produced sandalwood nuts, sugarcane, chickens, yams, and bananas.

Half a century later, Captain Cook, on his second voyage around the world, found an even less inviting scene. Only three or four canoes could be seen on the whole island. "These were very narrow, built of many pieces sewn together with fine line. . . . As small and mean as these canoes were, it was a matter of wonder to us where they got the wood to build them with; for in one of them was a board six or eight feet long, fourteen inches broad at one end, and eight at the other; where I did not see a stick on the island which would have made a board half this size." Cook also mentions the gigantic statues erected on stone platforms and several that had been toppled onto their faces. The account of Captain Cook's visit gives the impression that the island had just suffered a serious major conflict. Straw houses had recently been burned, and many fires lit the skies at night.

A few years later, the French mariner Jean François de Galaup, Comte de La Pérouse, spent just 8 to 10 hours on the island, but in this short time he identified the reason for its poverty of natural resources. The inhabitants, he said, had been so imprudent as to cut down all the trees that had grown there in former times, leaving it fully exposed to the rays of the sun and the sweep of the winds. Trees do not grow again in such a situation, he observed, unless they are sheltered from the sea winds, either by other trees or an enclosure of walls. La Pérouse attempted to help the islanders amplify their small food supply. He gave them lambs, hogs, and goats, as well as many seeds and tubers to start gardening. But the people were so improvident that they ate the animals before there was a chance for them to breed. The people also were not willing to work at building protective walls for the young trees or cultivating vegetables.

In 1872, a decade after the abduction of slaves by the Peruvians, the young Frenchman Pierre Loti was a crew member on a frigate that stopped at Easter Island. He was deeply moved by the tragic character of this tiny spit of land with its dying remnant of humanity. Loti, who later became a prominent writer, felt instinctively that there was something sinister about the island. "The country seems an immense ossuary," he said. "Skulls and jawbones we find everywhere. It seems impossible to scratch the ground without stirring those human remains."

By the time of Loti's visit all the completed statues had been toppled. They lay on their faces where once they had stood proudly dominating the landscape. Only the unfinished moai where the carvers had thrown down their tools two centuries earlier still looked up at the sky with sightless eyes. They lay in disarray, partially covered with brown grasses on the slopes of Rano Raraku.

The tragic history of Easter Island demonstrates a danger that is especially characteristic of island communities. Because they are isolated, there is no ready escape from a problem that threatens death or destruction. When food became scarce, the people of Easter Island could not move on to a more fertile environment. There was no place to go. Thousands of miles of treacherous waters separated their island from other lands. Because they had cut down their forests, there were no trees to build serviceable canoes, even if they had been willing to undertake such a perilous journey.

Like Easter Island, the Earth is a little island isolated in the vast seas of space. Its resources, although rich and various, are limited. And the increasing population of mankind is beginning to press hard upon these limits. Forests all over the world are being cut down to make space for growing more food and to provide fuel to cook the food. Thomas Lovejoy of the Smithsonian Institution estimated in 1989 that the world's tropical forests are being destroyed at the rate of 100 acres per minute (or 52 million acres a year). As resources become scarce, wars are fought over the division of the remaining supplies. Starvation is a fact of life in many human communities and violence is increasing around the world.

But perhaps the most alarming part of the message that comes down to us from Easter Island is the way violence breeds more violence. Acts of cruelty become progressively easier to commit when they are reinforced by example and supported by tradition. On the other hand, acts of kindness and compassion are reinforced in a civilized society. Human nature is complex, volatile, and impressionable. Capable of both good and evil, it can be molded by life experiences. An education in violence uncovers the beast in the nature of man.

We are faced with a critical challenge—to achieve a proper balance between the supplies of food and the needs of our people. We must find ways of stabilizing the human population and ways of increasing and distributing the produce of the world. This challenge must be met if the terrible last phases of the tragedy on Easter Island will not be repeated on planet Earth.

Chapter 3

THE GALÁPAGOS ARCHIPELAGO

Isles of Discovery

Observe always that everything is the result of change, and get used to thinking that there is nothing Nature loves so well as to change existing forms and to make new ones like them.

—MARCUS AURELIUS

In the Galápagos Archipelago nature has set up a tiny microcosm where we can see the principles on which life constructs new forms and how these forms interact and change through space and time. The islands, 600 miles off the coast of Ecuador, are very young, very small, and very strange. Certainly not beautiful, they are raw and unformed, dark lumps of matter, like scabs on the surface of the sea. Far beneath them a deep rift in the Earth's crust bleeds and swells. Like the Mid-Atlantic Rift, the Galápagos Rift emits springs of hot, sulfurous water and ejections of magma from the soft inside of the planet. Thus, the rift built layer after layer of lava and seamounts that have become islands.

Just a few miles from the Galápagos Islands is one of the most fractured places on the planet's surface. Known as the Galápagos Triple Junction, it is the point where three of the Earth's plates are pulling apart from each other: the Cocos, Nazca, and Pacific plates. The separation is so violent that it has created a very deep hole in the ocean crust—the 18,000-foot rift valley called the Hess Deep, which lies just north of the Galápagos Islands—and new crustal material is constantly forming at the spreading center.

Current geologic activity in the islands is intense. Earthquakes are common and volcanic eruptions are perhaps more frequent than anywhere else on Earth. In 1968 the collapse of the volcanic mountain on Fernandina Island creating a caldera was one of the largest such events recorded in history.

In more recent years, the fissure that gave birth to these pieces of the Earth's surface has been explored by research teams using submersible craft. The first observers (the Galápagos Hydrothermal Expedition in 1977) lowered cameras and thermometers toward the floor of the Pacific Ocean. At first, the temperatures recorded were near freezing. Then, suddenly, the temperature chart showed a sharp spike, and the vessel was taken closer to examine this remarkable phenomenon at close range. Through the portholes the diving crew looked out on a bizarre oasis of life and warmth set in the darkness of the abyss. The lights of the vessel revealed fountains of milky blue water pouring from fissures in the ocean bottom. The water was warm and laden with bacteria and tiny particles of sulfur. The rocks bathed by these warm springs were almost solidly encrusted with mussels and clams as large as a man's head. Armies of crabs scurried along the seafloor, feasting on sea worms and other small organisms that grew attached to the rocks. An octopus with fins like large bat ears floated by the window, and a fish swam by, waving a thin hairy tail. A few miles farther along the rift other hot springs were discovered. Each had its own characteristic community of living organisms. Huge tube worms monopolized one vent; their feathery plumes, swaying in the gentle currents, created a veritable fairyland.

Subsequent expeditions found that the islands rise from a shallow submarine platform. Although the surrounding seas are mostly 9000 feet or deeper, waters beneath the Galápagos Islands are, in general, less than 3000 feet deep. The oldest islands—Española, Santa Fe, Baltra, and Seymour—were formed by the uplift of lava flows originally spilled out on the ocean floor about 2.5 to 5 million years ago. The other islands are composed of younger volcanoes that erupted less than one million years ago. Geologically speaking, these are very young indeed. Wind and rain and sun have not had sufficient time to soften the harsh outlines of solidified lava flows.

The climate of the archipelago is almost as unusual as the geology of the seafloor underlying the islands. Although it lies on the equator, it is not hot and sultry year-round with abundant rainfall like many equatorial lands. In fact, the rainfall in the Galápagos is very sparse. And because it is bathed by cold waters brought by the Humboldt Current that flows northward from the Antarctic along the coast of South America, the surface temperature of the water that surrounds the islands varies between 63 and 72 degrees Fahrenheit during most of the year.

Another major current brings warm tropical waters from the Gulf of Panama. These two currents usually converge north of the Galápagos, but from January to April, the convergence may shift south, and the surface temperature of the waters that bathe the islands is considerably warmer (73–81 degrees Fahrenheit), and the air temperatures may reach 85 degrees. The warmth of this current, known as El Niño (named for the Christ child because it frequently affects the islands around Christmas time), alters the airflow patterns and has recently become famous for its significant effect on the climate of the United States, even of the whole world.

Even in the "rainy" season, when El Niño dominates, rain falls only sporadically, with otherwise clear skies. In the cool season there is little actual rainfall along the coastal regions—only fog, mist, drizzle, and a perpetually murky sky. The hills, which are actually volcanic mounts, are almost continuously enveloped in clouds and some rain does fall there.

Green vegetation is very sparse along the shores. Tangled thickets of wiry bushes grow in deep fissures of volcanic rock by parched growths of distorted cactus trees. Large specimens of Opuntia cactus dominate the landscape. There are Palo Santo trees, leafless and dead-looking in the dry season with pale trunks and white twisted branches. On higher ground, sedges and bracken ferns provide a more solid ground cover.

Centuries ago giant tortoises lumbered throughout this desolate landscape, looking for food in the sparse vegetation. These enormous, slow-moving reptiles, weighing up to 500 pounds, were present in vast numbers on the islands and provided an almost unlimited supply of fresh meat to reprovision the sailing ships that plied these waters. Thus, the islands were named "Galápagos"—the Spanish word for giant tortoises. As a result of the slaughter of tortoises for food, their population is sadly diminished; a surviving group is protected in the Tortoise Reserve and at the Charles Darwin Station at Santa Cruz Island.

Despite the tortoises' fate, the archipelago is not devoid of life. Like the scenes in the hot springs far below, it is an oasis of living things. The intermingling of the warm and cold ocean currents sets up many micro-habitats suitable for a wide diversity of species—cold water creatures like penguins, sea lions, and fur seals and warmth-loving flora and fauna like lizards and cacti and land tortoises. The rocky coastlines of all the islands are virtually encrusted with bizarre forms of life. Enormous iguanas and sea lions lie stretched out on the rocks, soaking up the warmth of the sunshine. Red Sally Lightfoot crabs are in constant motion, and at times the whole beach seems to be moving. On some islands blue-footed boobies, terns, and penguins live and breed so tightly packed together that they alter the whole texture of the land. Many of the life forms seen here can be seen nowhere else in the world.

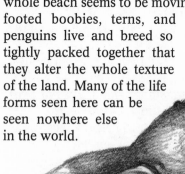

Like variations on a musical theme, familiar species have radiated into many different versions of the basic form. There are lizards that feed on seaweed beneath the ocean surface, sunflowers and cacti that have grown into trees, gulls that forage at night, and birds that cannot fly.

The unusual geological setting of these islands was unknown to Charles Darwin when in 1835, at the age of 26, he surveyed this extraordinary scene. But he sensed the singular nature of these tiny islands and was profoundly impressed. He carried the memory of the scenes with him on his cruise around the world. Long, uneventful days at sea led to thoughts that had a lasting impact on our understanding of nature and even of ourselves. Darwin wrote:

> The natural history of these islands is eminently curious and well deserves attention. Most of the organic productions are aboriginal creations, found nowhere else; there is even a difference between the inhabitants of the different islands; yet all show a marked relationship with those of America, though separated from that continent by an open space of ocean, between 500 and 600 miles in width. The archipelago is a little world within itself, or rather a satellite attached to America, whence it has derived a few stray colonists, and has received the general character of its indigenous productions. Considering the small size of these islands, we feel the more astonished at the number of their aboriginal beings, and at their confined range. Seeing every height crowned with its crater, and the boundaries of most of the lava-streams still distinct, we are led to believe that within a period, geologically recent, the unbroken ocean was here spread out. Hence, both in space and time, we seem to be brought somewhat near to that great fact—that mystery of mysteries— the first appearance of new beings on this earth.

Darwin observed that the species on each island are slightly different from the closely related ones on neighboring islands. Finches were the species that especially attracted his attention, about which he wrote:

> [There is] a most singular group of finches, related to each other in the structure of their beaks, short tails, form of body, and plumage; there are thirteen species. All these species are peculiar to this archipelago; and so is the whole group. . . . Two species may be often seen climbing about the flowers of the great cactus-trees; but all the other species of this group of finches, mingled together in flocks, feed on the dry and sterile ground of the lower districts. The males of all, or certainly of the greater number, are jet black, and the females (with perhaps one or two exceptions) are brown. The most curious fact is the perfect gradation in the size of the beaks in the different species of Geospiza, from one as large as that of a hawfinch to that of chaffinch and even to that of a warbler. The beak of Cactornis is somewhat like that of a starling; and that of the fourth subgroup, Camarhynchus, is slightly parrot-shaped. Seeing this gradation and diversity of structure in one small, intimately related group of birds, one might really fancy that from an original paucity of birds in this archipelago, one species had been taken and modified for different ends. . . .
>
> I never dreamed that islands, about fifty or sixty miles apart, and most of them in sight of each other, formed of precisely the same rocks, placed under a quite similar climate, rising to a nearly equal height, would have been differently tenanted; but ... this is the case. It is the fate of voyagers, no sooner to discover what is most interesting in any locality, than they are hurried from it; but I ought, perhaps, to be thankful that I obtained sufficient

material to establish this most remarkable fact in the distribution of organic beings.

The scene today is almost as extraordinary as it was in Darwin's time, except for the reduced populatior 3 of giant tortoises and of fur seals, which were hunted for their valuable skins. Feral goats and rats that undoubtedly found their way to the islands on ships have made inroads into the bird populations, especially those that nest on the bare ground. However, the islands still present a living, everchanging display where nature can be observed. Because of their unusual setting and isolation from human habitation, the Galápagos provide a unique opportunity for the study of natural history; it is a wonderland for nature lovers.

The animals are not frightened of human beings, and one can walk among them observing even the most intimate phases of their lives: courtship, mating, nesting, and the care and feeding of their young. One of the first rules for visitors to this wildlife sanctuary is "Do not touch the animals," and this is one of the hardest rules to follow. The temptations are so great. To see a baby blue-footed boobie—a soft ball of white down looking like an enchanting children's toy—and not to be able to pick it up is hard indeed, just as it is to look into the gentle brown eyes of a juvenile fur seal and resist stroking its shining smooth-fitting coat, tactile as wet velvet.

On Daphne Island I visited a crater floor where there was a nesting ground for boobies. The site was densely occupied with nests, casually thrown together with just a few sticks and leaves, every few feet apart on the bare ground. Threading my way carefully between them, I passed close to a pair of red-footed boobies engaged in a mating ritual—a precisely orchestrated movement with first the male and then the female stretching their heads up, pointing their bills toward the sky and revealing the dark triangular marking on the underside of the neck. When I passed back a short time later, I found the pair mating—wings spread wide.

I was amazed to observe the care and feeding of the young boobies—the almost desperate haste with which the parents flew endlessly back and forth to the sea, to catch more fish to fill the gaping mouths of their young. Many of these juveniles were as large as their parents, but they were still totally dependent. They stood in the nest site squawking for more food hour after hour.

On Tower Island there was a colony of magnificent frigate birds, whose enormous wing spread is wider in proportion to its weight than any other bird. I was sitting on a rock, holding a camera in my hands, waiting for an opportunity to photograph a male frigate bird inflating its mating signal—a huge red balloon on its neck. But suddenly one of these large birds walked right up to me and, disregarding the "don't touch the animals" rule, rested the knobby end of its beak right on my arm, so it could look more closely at the camera I held in my hands.

One of my most memorable experiences was a swim in a shaded tidal pool, well back from the beach, where a blowhole geyser was shooting up fountains of water every few minutes, making rainbows in the misty air. Nearby, in a lava grotto beneath a grove of Opuntia cactus trees, a small colony of fur seals were disporting themselves, diving into the blue-green pool, which was deep but so clear that I could see the bottom and the sides, which were smooth, with no sharp lava projections. It was a hot day and this seemed like a perfect place for a swim. I dove into the cool water, being careful not to encroach on the end where the fur seals played. They did not seem to be disturbed by my presence but continued their diving and swimming, nuzzling each other and making a soft barking sound like the hoarse lowing of a cow separated from her calf. These seals are said to be related to the fur seals that live in the icy waters of the Antarctic, carried here by the Humboldt Current.

On another day, when we went ashore on a sand beach at the north end of James Bay, I took along my mask and snorkel in order to get a glimpse of the underwater life. I did not see many fish near the shore, but as I swam over a deep hole in the ocean bottom, I saw swarms of angelfish, wrasse, and starfish. They were large and very colorful, different in markings from any I had seen before. I was told later that 23 percent of the shore fish here are found nowhere else in the world. Because the water was quite cool, however, I was not surprised to see that there was very little coral growth.

Most of the Galápagos animals have radiated into many subspecies. There are three types of tortoise, distinguished by the shape of their carapaces (upper shells), and there are variations within these groups. These differences are at least in part explained as adaptations to different climates and food supply. The saddle-backed species generally inhabit the smaller, lower, drier islands without much vegetation; their shell shape and neck and limb lengths allow them a greater reach when foraging for plants. This form, however, has its disadvantages; if the vegetation is low and dense, the raised shell in front becomes easily caught on branches and vines. Thus, the dome-shaped tortoise and related species are found generally on well-vegetated highlands, where a long reach is not necessary to obtain food. On the island of Isabela there are five distinct types of tortoise.

The iguanas on the Galápagos are also very diverse. There are two forms of iguanas—large, sluggish, and yellow-colored lizards with a row of spines on their backs, like small Stegosaurus dinosaurs. Marine iguanas are not as large and heavy, but they are black, "hideous-looking creatures," as Darwin described them. This is the only lizard in the world that lives in an intertidal zone and feeds primarily on marine algae. Their short, blunt noses allow them to crop the close-growing seaweed from the rocks exposed at low tide. They also dive to feed on seaweed on the ocean floor. They have been observed at a depth of 35 feet and can stay under water for at least an hour. There are seven subspecies of marine iguanas showing variations in size and color from island to island.

About two-thirds of the resident birds in the islands are endemic. Beside Darwin's 13 species of finches, there are three species of boobies, three of storm petrels, and four of mockingbirds. Nature seems to be profligate in creating so many variations on every theme. The tempo of speciation is especially apparent on islands, because there are many habitats available (not already occupied), and the species that settle down in each of these territories adapt to their particular requirements and opportunities. Over many generations they evolve characteristics that give them biological advantages.

As Darwin observed, the various species of ground finches found on the different islands have adapted to small variations in the food supply. A heavy beak is useful for grinding hard seeds; a slender beak is more suitable for impaling small insects and picking out the soft tissue of the Opuntia cactus. Because predators were not a significant factor, many of the subspecies were able to find an environment where they could thrive and multiply.

It is not surprising that some new forms of life fall between the two extremes—the successful adaptations and those that do not survive. In an island situation, these life forms are frequently able to find a niche where they can live and breed with reasonable success. Sometimes these species are known as "relicts" (implying that

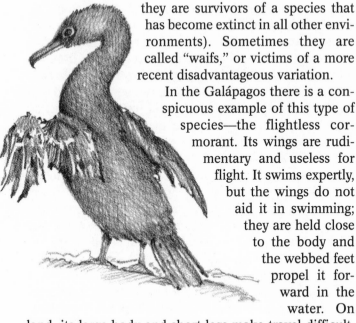

they are survivors of a species that has become extinct in all other environments). Sometimes they are called "waifs," or victims of a more recent disadvantageous variation.

In the Galápagos there is a conspicuous example of this type of species—the flightless cormorant. Its wings are rudimentary and useless for flight. It swims expertly, but the wings do not aid it in swimming; they are held close to the body and the webbed feet propel it forward in the water. On land, its large body and short legs make travel difficult. Rough terrain is traversed by a series of springy jumps from rock to rock, the tiny wings being held out for balance. After swimming the cormorant climbs to an elevated location where it extends its wings to dry in the breeze. Seeing it thus, one is impressed by the pitiful appearance of these wings, like two black ragged shirts pinned up on a line to dry. This creature does indeed fit the mental image of a waif.

The flightless cormorant is classified as rare; the total population was estimated in 1972 as 800 pairs, and it is found only on the coasts of Isabela and Fernandina. For its food supply, it needs cold shallow seas with rocky bottoms where it can feed on eels, octopuses, and fish. Off the shore of these two islands, it has found small regions where these conditions are met.

Darwin recognized that many of the animals that inhabited the Galápagos were different although similar to the usual form of that animal that he had seen elsewhere. The variations were distinct, making separate species or at least subspecies. This was a surprising discovery and quite disturbing to him. Darwin and most of the educated men of his time had been brought up on the doctrine known as the *Scala naturae* or Chain of Being. All of nature, from minerals up through the lower forms of life to man, was believed to be arranged in a ladder of increasing complexity. But this ladder was static. There was no suggestion that a lower order of complexity could evolve into a higher one. Each species was the result of a separate act of creation. And creation was not considered to be still in progress. Because both theologians and scientists of the time believed the world to be only a few thousand years old, there was literally not enough time for evolution.

There was, however, a new stream of philosophical thought that was beginning to suggest—although very cautiously—that the Chain of Being was not completely static. The eighteenth-century French naturalist Comte de Buffon described in great detail many oddities of nature, and he ventured warily that among the numerous families brought into existence by the Almighty, "there are lesser families conceived by Nature and produced by Time." He also recognized the need for a much longer time scale in order to account for the geology of the planet and the history of life upon it.

Darwin's grandfather, Erasmus Darwin, wrote a two-volume treatise entitled *Zoonamia*, published in 1794 and 1796, in which he estimated the antiquity of the Earth in millions of ages and said that despite the diversity of living things he believed that "the whole is one family of one parent." In another work, he carried this thought further: "As all the families both of plants and animals appear in a state of perpetual improvement or degeneracy, it becomes a subject of importance to detect the causes of these mutations." His explanation of the mechanisms of evolution lay in "the power of acquiring new parts, attended with new propensities, directed by irritations, sensations, volitions, and associations; and thus possessing the faculty of continuing to improve by its own inherent activity, and of delivering down those improvements by generation to its posterity, world without end!" The very revolutionary thought that time has played an essential role in producing the diversity of nature is clearly expressed here.

Jean-Baptiste Lamarck, another French naturalist, published a number of works in which he expounded his own theory of evolution. He believed that all change "was the accomplishment of an immanent purpose to perfect the creation." Thus, the old fixed ladder of being was transformed into an escalator. Life is constantly emerging, remaking itself through its own inner perfecting drive; it moves toward greater complexity and ascends toward higher levels.

Charles Darwin had been exposed to these philosophies, but he was also aware that these views ran counter to the generally accepted theological explanations of the history of life. And even if one could accept the general principle of evolution, there was the question of how changes occurred—the mechanism of organic modification. Lamarck and Erasmus Darwin invoked an ancient hypothesis that had widespread support in the folk belief of the time—the inheritance of acquired characteristics.

Charles Darwin did not accept these explanations. His observations while on the survey voyage of the *Beagle* (1831–1836) suggested to him quite a different explanation. He had recently read Thomas Malthus's *An Essay on the Principle of Population* (1798), which pointed out that people multiply more rapidly than the quantity of food increases. Thus, many more individuals are born than can possibly survive. The competitive pressure favors the most able. Having observed the struggle for existence that goes on everywhere in nature, it seemed reasonable to Darwin to assume that the same principle could be applied to all species. Favorable modifications would tend to be preserved while unfavorable ones would be destroyed. Thus, long-term change—evolution—could be based on natural selection.

The idea was not fully formulated until many years later, but it gestated in Darwin's mind after observing the unusual distribution, the exotic modifications, and the dense populations of species on the Galápagos Islands. But none of these observations explained why modifications occurred in the first place. An important piece of the evolution puzzle was missing. The report of the significant experiments on pea plants conducted in a monastery garden by Gregor Johann Mendel and published in an obscure journal in 1866 was not discovered and appreciated by the scientific world until the early years of the twentieth century. Mendel showed that inheritance is carried by small biological units and that these are remarkably stable generation after generation. These units—later named genes and chromosomes—were subsequently found to occasionally undergo mutation, events that appeared to occur in a random manner.

This body of scientific knowledge supported Darwin's theory of natural selection. Nature does tend to create variations on existing forms of life throughout time. Some of these are advantageous, many are deleterious and do not survive, but others are relatively neutral—good in some environments but less favorable in others. In an island situation—and the Galápagos are a prime example—there are many habitats available, so a new subspecies can find an environment where their small differences are advantageous. Natural selection working on these chance variations has resulted in the evolution of life forms of wonderful diversity and increasing complexity. This is basically (with some minor modifications) the accepted explanation today.

In Darwin's theory, as well as those of Buffon and Lamarck, time plays a crucial role. Both the biological world and the physical one are constantly undergoing change, establishing greater variety and complexity in life forms. It is an interesting observation that human beings have been very slow to acknowledge the reality of change. We seem to have been more comfortable and secure in the belief that everything is just the same as it was in the beginning, and always will be. If many things that happen can have a lasting effect on our world, then the future is open, unknown, and a little frightening. It has been a very slow intellectual process of increasing knowledge and understanding that has made it possible to overcome this fear, to admit the reality of change and progress. The Darwinian perception played an important role in bringing about this new worldview. It is fascinating to think that these revelations about the history of life and the planet itself were first conceived on a trip through the Galápagos—those relatively unformed islands that had sprung forth from the womb of the Earth in the last few million years, the tiniest fraction of geologic time.

Now we recognize that "all phenomena have a historical aspect," as the twentieth-century biologist Julian Huxley said, "from the condensation of nebulae to the development of the infant in the womb, from the formation of the earth as a planet to the making of a political decision, they are all processes in time; and they all are interrelated as partial processes within the single universal process of reality. All reality, in fact, is evolution, in the perfectly proper sense that it is a one-way process in time; unitary; continuous; irreversible; self-transforming; and generating variety and novelty during its transformations."

This understanding has changed our way of looking at the universe. "What makes and classifies a 'modern' man," the French biologist Pierre Teilhard de Chardin wrote, "is having become capable of seeing in terms not of space and time alone, but also of duration . . . and incapable of seeing anything otherwise—anything—not even himself."

Chapter 4

BALI AND LOMBOK

Wallace's Line

Truth is born into this world only with pangs and tribulations, and every fresh truth is received unwillingly. To expect the world to receive a new truth, or even an old truth, without challenging it, is to look for one of those miracles which do not occur.

—ALFRED RUSSEL WALLACE

Much has been written about Charles Darwin and the Galápagos Islands, and there is a general impression that the theory of natural selection could not have been conceived anywhere else on Earth. The truth is that many places—especially islands—could have served as the spark that illuminated the mystery of the origin of species. In fact, another Victorian naturalist came to the same conclusion on the opposite side of the globe at nearly the same moment in history.

Alfred Russel Wallace, an Englishman fourteen years younger than Darwin, had been traveling in the Far East since 1854, studying the flora and fauna and especially noting the differences in distribution of the species. He had been turning over in his mind this important question: What causes the appearance of new species? He had reached Ternate in the Malay Archipelago when the answer burst fully formed into his mind—the principle of the survival of the fittest. Wallace had read Thomas Malthus's *Essay on the Principle of Population*, which drew attention to the competition for existence that acts as a brake on human population, which increases more rapidly than the food supply. Under this competitive pressure, favorable variations tend to be preserved and the unfavorable ones to be destroyed. Wallace saw that this same principle could be applied to the flora and fauna and could account for the formation of new species.

Wallace sat down and wrote out these conclusions. He sent the article, which he had entitled "On the Tendency of Varieties to Depart Indefinitely from the Original Type," to Charles Darwin, with whom he had corresponded earlier about his observations of nature in many out-of-the-way places around the world. In the meantime, Darwin, who had also read and been influenced by Malthus, had been working on the same theory, which had occurred to him almost twenty years earlier. In March 1858 he was startled to receive in the mail the essay written by this relatively unknown young naturalist. Wallace's essay stated in a very concise and convincing manner the central concept of the theory of natural selection.

The unexpected competition from Wallace spurred Darwin into rushing his own work to completion, but, at the same time, he did respect Wallace's rights to credit for the discovery. Darwin's friends Charles Lyell and Joseph Hooker arranged for a simultaneous presentation of his theory and Wallace's at a meeting of the Linnaean Society in August 1858 and for the publication of Wallace's paper in the *Journal of the Proceedings of the Linnaean Society*.

Darwin continued to vigorously support the theory of evolution by natural selection, which shocked the British public. They had always believed that man was created in God's image, and now they were told that he was descended from an ape! In 1859 Darwin published *On the Origin of Species by Means of Natural Selection*, his detailed exposition of the theory. While Darwin is remembered for this explanation of the appearance of new species on Earth, Wallace is generally forgotten. But Wallace made another series of observations that were of great significance to evolutionary biologists, and his name has always been associated with this perception.

Wallace said:

I have arrived at the conclusion that we can draw a line among the islands [of the Malay Archipelago], which shall so divide them that one-half shall truly belong to Asia, while the other shall no less certainly be allied to Australia. I term these respectively the Indo-Malayan, and the Austro-Malayan divisions of the Archipelago. . . . The great contrast between the two divisions of the Archipelago is nowhere so abruptly exhibited as on passing from the island of

Bali to that of Lombok, where the two regions are in closest proximity. . . . The strait is here fifteen miles wide, so that we may pass in two hours from one great division of the earth to another, differing as essentially in their animal life as Europe does from America.

This division is now called Wallace's Line and is clearly marked on most maps of that area.

—

The island of Bali is one of the most exotic and perhaps the most alluring of all the islands in the world, so it may be surprising to learn that it has only been an island for a relatively short period of geologic time—perhaps 12,000 years. It is separated by a shallow, narrow strait from Java, which is separated by another strait from Sumatra, the Malay Peninsula, and the continent of Asia. Bali has been part of this larger land mass at periods of low sea level, when glaciers imprisoned great volumes of the Earth's seawater. Therefore, Bali is a continental island and has not been severely isolated—not like very old islands that have moved decisively away from their parent continent, such as Madagascar or the Seychelles, and not like the new islands of Kauai or the Galápagos, which have recently erupted from the bottom of the sea. It also is not surprising to find that many of the distinguishing marks of island ecology (large numbers of endemic species, high rates of extinction) are not present in Bali. But like many islands, it does have great natural beauty—a mountainous landscape bordered by black volcanic and white coral beaches and lapped by the warm waters of the Indian Ocean.

Because it was quite recently a part of Asia—during the lifetime of mankind— Bali took on many of the cultural characteristics of that land, and these traits are still evident. It is a cultivated island where human beings have lived and worked for many centuries. Most of the land is divided up into small, irregular patches and is carefully tended. On the steep mountainsides many terraces support rice fields that can be flooded or drained by means of a system of irrigation ditches and small channels to catch the streams that flow down from the higher elevations. Even in the middle of the nineteenth century, when Wallace visited Bali, these agricultural practices were well established. "I had never beheld so beautiful and well cultivated a district out of Europe," he said. Today, they are still farmed in the same way. On the precipitous hillsides the rice paddies make hanging tapestries woven in the most exquisite shades of green.

The climate of Bali is benign, lying just 8 degrees below the equator in the belt of the trade winds. And because the land was volcanic, it is very fertile. The volcano Agung, which had been quiescent for 120 years, erupted in 1963, killing 1500 people and leaving thousands homeless. But as the ash weathered, it produced a rich, friable soil favorable for farming. Bali produces three crops of rice a year. The only difficulty for agriculture is the steep contour of the land, but this has been overcome by arduous construction of terraces.

The narrow main road that twists precariously up the mountainside is decorated at almost every turn with a Hindu temple, constructed of dark red bricks made from the local clay, that are lavishly ornamented with stone sculptures. The Oriental love of intricate pattern and extravagant design is displayed in each of the little communities that are strung out along the highway. Every village has a gate with red brick pillars and stone sculptures to mark its entrance, and the houses that line the road are like small versions of the temples—everywhere dark-colored, intricately designed structures. The whole looks as though it might have sprung organically from the mountainsides.

The people of Bali are artistic and mystical. They have carried the art of carving, of both wood and stone, to a high level of achievement. On a smaller scale, they specialize in filigree metalwork, using gold and silver for jewelry and other decorative items. Their lives, too, are decorated with frequent festivals and ceremonial days. Balinese culture is centered on religion—a blend of Hinduism, Buddhism, and animism with magical beliefs and practices. One very unusual ceremony was described by Lawrence Blair in his book *Ring of Fire*:

> A Balinese child does not touch the ground for the first three months of life. He is cradled and cosseted above the earth and introduced to gravity very gently. When he is 105 days old a "foot-touching-the-ground" ceremony is held, when the child is ritually "planted" in matter, and first sets foot on the earth. Until then he has merely been an angel, hovering at the frontiers of the heavenly world. . . . A Balinese, like a tree, must remember that he is strung between two worlds, balanced between the pull of gravity and the pull of heaven.

Most of the flora and fauna on Bali are closely related to the species that are found on Java. Domestic cattle descended from the banteng (wild ox) of Java are raised here, and, in Wallace's time, wild cattle of the same species were said to be found in the mountains. Two kinds of monkeys and one species of tiger also roamed these tropical landscapes, rich with tamarind, fruit trees, and coconut palms.

But it was the geographical distribution of birds that seemed especially significant to Wallace. In Bali he found a weaverbird with a bright yellow head that builds its bottle-shaped nests in trees near the beaches. He also found a number of other species he knew to be native to Java, including a wagtail thrush and several starlings. One of the starlings is a gorgeous bird with gleaming white plumage and a mask of turquoise. But Wallace was very surprised to find when he crossed the narrow strait to the island of Lombok to the east that none of these birds was present on that island. Instead, there were honeysuckers, large green pigeons, and brush turkeys, which were equally unknown in Bali or any island

farther west. Bali and Lombok, he declared, differ as essentially in their animal life as Europe does from America.

One reason for this striking difference in animal life, Wallace surmised, is the depth of the strait that separates these islands. The Lombok Strait, although only 15 miles wide, is very deep. So probably even at times of low sealevel an ocean passage lay between Lombok and Bali. The importance of this barrier was reinforced in Wallace's mind by his experience when he crossed the narrow strait. Wallace described violent surf on the beaches and the bays overlooking this strait:

> Sometimes this surf increases suddenly during perfect calms, to as great a force and fury as when a gale of wind is blowing, beating to pieces all boats that may not have been hauled sufficiently high upon the beach, and carrying away incautious natives. This violent surf is probably in some way

dependent on the swell of the great southern ocean, and the violent currents that flow through the Straits of Lombok. These are so uncertain that vessels preparing to anchor in the bay are sometimes suddenly swept away into the straits, and are not able to get back again for a fortnight. What seamen call the "ripples" are also very violent in the straits, the sea appearing to boil and foam and dance like the rapids below a cataract; vessels are swept about helpless, and small ones are occasionally swamped in the finest weather and under the brightest skies.

I sailed across this strait in a catamaran cruise ship and was warned that the ocean might be very rough for this passage. Fortunately, that night it was unusually calm. At dawn the next morning from shipboard, I saw the two islands—so similar in size and shape—their twin volcanoes, Agung and Rindjani, rising out of the misty sea to more than 10,000 feet and glowing gold as they caught the first rays of the rising sun.

Lombok, like Bali, is clothed with tropical vegetation, and dark beaches of volcanic sand edge the clear blue sea. As in Bali, many square miles of rice terraces have been built and irrigated by channels, so every portion of them can be wet or dry, depending on the needs of the crop. Women do most of the fieldwork by hand. They also weave beautiful patterns into cloths of long staple cotton, and these ikats are worn in various ways, forming the usual clothing of both men and women.

The lifestyle of the people of Lombok is characterized by a strong sense of community. One day I visited a very remote mountain village with small houses built of bamboo and thatch with raised floors and one open side fronting on a large square. Several of the buildings appeared to have a common function. In one, very small children were playing together and babies were tended. In another, food was being prepared, perhaps more than usual that day because a special ceremony was scheduled to take place—the circumcision of all the little boys ages two to five. This was an occasion for great celebration. In the village there seemed to be an almost unlimited supply of older children who did not beg but followed us as though we were the Pied Piper of Hamelin, trying out their few English words on us.

As in Bali, religion and ceremony play a very important part in these people's lives. On Lombok, however, the majority of the inhabitants—about 85 percent—are Muslim, while only about 10 percent are Hindu and 5 percent are Christian. But all three faiths are tinged with animism and traditional mysticism. Perhaps this common thread is the reason why there is an astonishing religious tolerance among the people of Lombok, as in most of the Indonesian islands. We visited one temple that is used by Muslims for their regular services but is also used by Hindus on occasion.

The people here are brown-skinned and handsome, similar in appearance to those on the Malay Peninsula. Human beings, of course, have a higher degree of mobility than most other fauna. Because they can build boats and sail back and forth with ease, they were not separated by Wallace's Line. Other animals, however, including the birds, are more closely related to Australian species, as Wallace observed:

> Birds were plentiful and very interesting, and I now saw for the first time many Australian forms that are quite absent from the islands westward. Small white cockatoos were abundant, and their loud screams, conspicuous white color, and pretty yellow crests, rendered them a very important feature in the landscape. . . . [There were] some small

honeysuckers of the genus Ptilotos and the strange
mound-maker (Megapodius gouldii), . . . a small
family of birds found only in Australia and the
surrounding islands. . . . Most of the species of
Megapodius rake and scratch together all kinds of
rubbish, dead leaves, sticks, stones, earth, rotten
wood, etc., till they form a large mound, often six
feet high and twelve feet across, in the middle of
which they bury their eggs. ... A number of birds
are said to join in making these mounds and lay
their eggs together, so that sometimes forty or fifty
may be found.

The eggs of these birds are left alone to be hatched
by the heat of the sun or of fermentation. Wallace was
especially entranced by the beautiful ground thrushes,
admiring their soft, puffy plumage and brilliant marking. He wrote, "The upper part is rich, soft green, the
head jet black, with a stripe of blue and brown over
each eye; at the base of the tail and on the shoulders
are bands of bright silvery blue, and the under side is
delicate buff, with a stripe of rich crimson, bordered
with black on the belly." He also admired "beautiful
grass-green doves, little crimson and black flower-
peckers, large black cuckoos, metallic king-crows,
golden orioles, and the fine jungle-cocks—the origin of
all our domestic breeds of poultry."

Although Wallace had surmised that the presence of
the Lombok Strait was the reason for the sharp dividing
line between the animal species of Bali and those of
Lombok, he did not seem to be disturbed by the fact that
only 15 miles divide the islands, a distance that most
birds can easily fly across. The ocean depth in the strait
would be no deterrent to a bird.

There are other reasons, however, why species may
find it difficult to colonize adjacent islands. A single bird
making the crossing cannot establish a new community.

Enough individuals must undertake the journey, and they must find a favorable ecological niche in which to survive and reproduce to make a viable population. In an island already settled, the niches may all be filled.

Biogeography also provides another clue to this phenomenon. Around the Earth there have been centers of biological development where the conditions are especially favorable and a dominant group of coadapted species has evolved. Eventually, they spread out from that center. In general, the least successful members of this biotic community move toward the periphery of the circle, where they do not need to compete directly with the more successful ones. If they find a place where they can live and breed effectively, they settle there and are not driven by the need to move on.

Large land masses are especially favorable for the evolution of these centers. They have more varied environments, and they accommodate more species than small land masses like islands. For example, the continents of Australia and Asia are ideal for the development of such centers, and we can imagine that islands like Bali and Lombok each lie on the periphery of one of these great centers.

In his travels throughout the Indonesian islands, Wallace found that the differences in Asian and Australian fauna were equally striking between Borneo and Celebes and all the islands eastward to Australia and New Guinea:

> In the first [on the west side of the line between Borneo and Celebes] the forests abound in monkeys of many kinds, wild-cats, deer, civets, and otters and numerous varieties of squirrel are constantly met with. In the latter [east of the line] none of these occur . . . no apes or monkeys, no cats or tigers, wolves, bears, or hyenas; no deer or antelopes, sheep or oxen; no elephant, horse, squirrel, or rabbit: none, in short, of those familiar types of quadruped which are met with in every other part of the world. Instead of these, it has Marsupials only, kangaroos and opossums, wombats and the duck-billed platypus. . . . All of these striking peculiarities are found also in those islands which form the Austro-Malayan division of the Archipelago.

Again, the reason for this important difference in the fauna lies in the nature of the ocean bottom surrounding these islands. Sumatra, Borneo, and Java, as well as Bali, lie on a continental shelf that, at times of low sea level, would have been dry land and continuously connected with the Malay Peninsula and Asia. The islands of Celebes and Lombok were separated from Asia by deeper seas. But farther east, a shallow sea lies between New Guinea and Australia, suggesting that these were also connected by dry land at times of low ocean level.

Wallace's Line runs between Bali and Lombok, then swings north and separates Borneo and Celebes. The area including Australia and New Guinea and bounded on the western side by Wallace's Line is like one very large island that was isolated for a long period of time from the rest of the world. Here living creatures evolved in their own special ways, resulting in very different species. Almost all the mammals are marsupials, producing their young in a neonatal state and then carrying them in a pouch, each one attached to a nipple and sheltered until fully developed. Many extraordinary creatures are members of this order—the kangaroo, koala bear, wombat, and bandicoot. There are marsupials that resemble bears, squirrels, wolves, moles, and even mice. Unusual birds also evolved in this isolated environment—the emu, cassowary, and the bird of paradise, a bird so beautiful it was believed to have come from heaven.

The author and world traveler David Quammen described one of these extraordinary birds, which he saw in the Aru Islands near New Guinea:

> It is a male of the . . . greater bird of paradise in full
> breeding plumage. The head is a deep satiny yellow;
> at the throat there's a patch of iridescent green; the
> breast, the back, and the wings are auburn; and
> from under the wings emerge long yellow plumes,
> which fan out behind like a ceremonial cloak.

It is strange to imagine that this exquisite creation evolved in the dark rain forests of New Guinea, molded by the same artist as the "hideous looking creatures"—the black iguanas of the Galápagos.

Chapter 5

THE ISLANDS
OF INDONESIA

Fire, Gold, and Spice

'Tis the spirit in which the gift is rich,

As the gifts of the wise ones were;

And we were not told whose gift was gold

Or whose was the gift of myrrh.

—Edward Vance Cooke

The islands of Indonesia are a remarkable part of the world, because they are the location of some of the most intense volcanic action on the planet. There is a chain of volcanoes running through the islands then up the coast of Asia through Japan, the Kamchatka Peninsula, and the Aleutian Islands to Alaska and down the west coast of the United States, Central America, and South America. Thus, the line of volcanic activity approximately circles the Pacific Ocean. This "ring of fire," as it is called, is caused by the forcing of the Pacific plate and several smaller plates under the continental plates of Asia and the Americas. As the ocean crust is thrust beneath the edges of the continental plates, it is heated, and hot magma erupts in volcanic action along the line where the plates meet and interact. Deep earthquakes are also common in these zones.

Several of the most violent volcanic eruptions recorded in human history have occurred in the Indonesian Archipelago, and they have killed more people than eruptions anywhere else on Earth.

About 700 miles northwest of Bali, the Strait of Sunda divides Java from Sumatra and at its narrowest point is 14 miles wide—similar to the Strait of Lombok. It also is very shallow, averaging about 600 feet. But it has been one of the great sea lanes of the world, carrying trade between Europe and the East Indies and China. At the western entrance to the strait, there is the island of Krakatoa, which once consisted of several volcanic cones. Prior to 1880, the island was not inhabited, and there was no record of volcanic activity. But in September of that year, a series of minor earthquakes began to shake the area around the Sunda Strait. This was not particularly alarming to the local people, because earthquakes were very common in this active seismic zone. However, the activity gradually increased until May 1883, when a volcano on Krakatoa came abruptly to life with a series of explosions that were heard 100 miles away. A great column of smoke rose at least 8 miles above the island, showering ash over points 300 miles away. By the end of May, the activity seemed to have quieted down to the point where it appeared to be safe enough to visit the island and see what was going on. An adventuresome group from Batavia (now Djakarta) chartered a boat for the excursion. They found that the island was covered ankle-deep in fine white ash, like snow. The trees on the northern part of the island had been stripped of their leaves and branches by a rain of pumice. The northernmost volcano was erupting almost continuously. Peering down into the crater, they saw that it was about 3000 feet in diameter and 150 feet deep. From a small opening in the crater floor a cloud of steam was erupting with a great roar.

The volcanic activity seemed to quiet down further, until the middle of June, when another eruption began in the center of the island. These eruptions dragged on for three months, until the climax was reached on August 26-27. The explosions began at 1 P.M. on the 26th and increased steadily, and by 5 P.M. they were so powerful that the sound could be heard all over Java. Ships that happened to be in the Strait of Sunda reported alarming effects that made navigation of the strait impossible. Captain Woolridge on the British ship *Sir Robert Sale* described the appearance of the column rising from the volcano. "It had a most terrible appearance," he said. "The dense mass of clouds being covered with a murky tinge, with fierce flashes of lightning." The whole scene was lit up from time to time by electrical discharges, and at one time the cloud above the mountain presented "the appearance of an immense pine tree, with the stem and branches formed with volcanic light." Peculiar pink-colored glows lit up the mastheads and rigging of the ships, because the atmosphere was highly charged with static electricity. This phenomenon, called St. Elmo's Fire, is generated by the rush of steam through the volcanic vent, and the friction between the fragmentary particles that were blasted up with it. The native crew of another ship were terrified that the light was the work of evil spirits, and they tried to put it out with their hands.

During the morning of August 27, Krakatoa exploded with such a force that the noise was heard over a large part of the Earth's surface. In south Australia, 2200 miles away, the noise was loud enough to wake sleeping people. It was even heard 3000 miles away at Rodriguez Island near Mauritius in the Indian Ocean. But the most frightening and destructive effects came from the sea.

Sudden alterations in ocean level caused by the changes in the shape of a volcanic cone (or by an earthquake) produce *tsunamis,* giant waves that move across the ocean with lightning speed. Velocities from 300 to 500 miles an hour in the open sea are characteristic. A tsunami is an unusual wave, not tremendously high as it moves through the ocean, but it consists of a series of very long oscillations, as much as several hundred miles from crest to crest. Ships at sea do not even notice their passing, so gradual is the rise and fall. But when the wave approaches a shelving shore or a harbor mouth, the water piles up and can do frightful damage in locations far removed from the original source.

On August 27, 1883, a series of tsunamis 100 feet high swept up and down the shores of the Sunda Strait, drowning low-lying areas all along the coasts. Several towns and villages simply ceased to exist as the great waves washed over them, carrying away the flimsy buildings. An estimated 36,000 people died.

The rain of ash and lumps of pumice from Krakatoa blotted out the sunlight. By 11 A.M. total darkness had fallen on Batavia, 100 miles from Krakatoa. When the skies finally cleared, it was found that two-thirds of the island had disappeared; a great crater had been formed, most of it below sea level; and where land had once stood 1000 feet above sea level, the water was now 1000 feet deep.

Krakatoa has sometimes been called the most violent eruption ever witnessed by humanity, but there is evidence that an eruption on Santorin in the Aegean Sea in the middle of the fifteenth century B.C. was several times more destructive. On April 15, 1815, on the island of Sumbawa, next door to Lombok on the eastern side of Wallace's Line, the mountain named Tamboro exploded in an event that lasted for five days. Afterward, the mountain, which had been 13,000 feet tall, was only 9000 feet, and a crater 12 square miles in diameter was formed. It is estimated that more than 40 cubic miles of debris was catapulted out of the mountain. Ash was deposited on everything within 600 miles of the volcano; Bali and Lombok must have been deeply covered in ash. About 10,000 people were killed instantly and 66,000 died later of starvation or disease. The debris thrown up into the atmosphere blocked up to one-fifth of the sun's light and heat, causing crop failures and unusual weather patterns worldwide. In the northeastern United States, snow and frost continued through June. In Great Britain and Scandinavia, there was almost continuous rain from May to October, resulting in poor harvests and food shortages, and the year 1815 was known as the year without a summer.

Volcanoes are terribly destructive of humans and the little communities they have set up around the world, but they are also the builders and makers of mountain ranges, fertile land, gemstones (emeralds, rubies, and diamonds), and of the very islands themselves. In the fiery furnaces of volcanic action, another creative process occurs, laying down treasure troves—deposits of many metals, including silver, copper, platinum, and gold.

Gold is widely distributed around the planet. Although it is present in all rocks, it represents only 1 part in 250 million of the continental crust. It is slightly more abundant in oceanic crust, and even in seawater, but the percentages are still extremely small. Rich veins of gold, however, are often found in the eroded roots of old volcanoes. In fact, it has been said that if the position of all the gold deposits formed within the past 60 million years were plotted, they would turn out to be scattered in a well-defined belt all around the Pacific, closely paralleling the ring of fire.

The explanation for this distribution—like the position of Wallace's Line—lies in the nature of the ocean crust. Around the perimeter of the Pacific and the adjacent seas, as oceanic crust is subducted, it becomes molten, and mineral-rich solutions percolate upward along paths opened up by the volcanoes. When temperatures and pressures are just right, gold begins to crystallize out of these concentrated solutions, and veins of pure gold are created.

Other treasures even more valuable were discovered in the tiny islands that lie scattered like pearls along the equator between the Malay Peninsula and Australia. These islands, known as the Moluccas or the Spice Islands, are a group of half-submerged volcanic mountains south of the

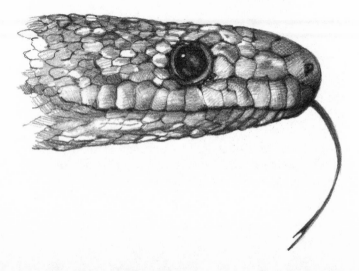

Philippines and west of New Guinea. On the tops of these isolated peaks certain very special evergreen trees evolved and flourished—endemic clove and nutmeg trees. Found nowhere else in the world, they belong to the angiosperm family, those remarkable plants that provide nutritious food for mankind as well as the colorful plants that have filled the Earth with beauty.

The nutmeg tree grows to about 65 feet and starts producing fruit at eight years of age. This is a doubly valuable tree because it produces two spices: mace, which is derived from the casing that surrounds the seed, and nutmeg, which is the seed itself. The clove tree is a slightly smaller plant, growing to 25 or 40 feet, and flowering begins a little earlier, about the fifth year.

It was discovered very early in human history that the seeds or flower buds of these trees contained highly flavored, aromatic spices that could add seasoning to our food. Although this may not seem to be of great importance in human affairs, at one time these spices were worth more than their weight in gold!

These and other spices have been used in cooking for so long that the custom predates recorded history. They were found to be effective as a preservative for meat and fish at a time when no refrigeration was available, and if the preservation was not totally effective, the flavor of a strong spice made slightly spoiled food at least palatable. Spices also provided cures for physical ailments centuries before scientific medications were available. They were used for embalming and for incense and perfume, because their powerful aromas were effective in masking other less pleasant odors. As early as 200 B.C., envoys from Java to the Han dynasty court of China brought along cloves, which they held in their mouths to perfume their breath during audiences with the emperor.

Several important spices were also found in India and Ceylon. Pepper was the most important of these for preserving meat and flavoring many types of food; it is derived from a vine that grows profusely in these tropical settings. Another is cinnamon, perhaps the all-time favorite spice. It also comes from an evergreen tree of the angiosperm family. Harvested in the wet season, the young shoots are cut close to the ground. Then they are stripped of their bark, which is wrapped into a quill 3 or more feet long and dried. Cinnamon has many applications. In Egypt it was used for embalming and witchcraft, in medieval Europe for religious rites and as a flavoring for a variety of foods, from confections to curries.

The most exotic of all seasonings is saffron. Its origin is unknown—perhaps Kashmir in northern India—but it now grows in many Middle Eastern and Mediterranean areas. Saffron is the sweet-smelling herb mentioned in the Song of Solomon; it was strewn in Greek and Roman halls, theaters, and baths as a perfume; and the streets of Rome were sprinkled with saffron when Nero made his entrance into the city. It was believed to have wonderful healing properties, as mentioned in an English reference book of the tenth century.

Saffron is derived from a purple-flowered crocus. Three stigmas are handpicked from each flower and dried over charcoal fires. It is widely valued as an unusual flavoring and as a fabric dye. Buddha's priests made saffron the official color for their robes. This yellow, as strong as distilled sunlight, heightens the visual impact of a thousand scenes of the Far East. And, because so little is produced from each flower, it is still the most expensive seasoning in the world.

The herbs frankincense and myrrh, obtained from Oriental trees, were also highly valued in ancient times as ingredients of perfume and incense. They diffuse a powerful fragrance when burned. Frankincense was used in medicinal plasters, too, while myrrh was used as an ointment, as a stimulating tonic, and even today as a mild antiseptic.

These spices and herbs from the Far East were light in weight, easy to transport, and deeply desired by the entire civilized world. They formed the basis of one of the most lucrative trades in human history. Arab merchants were the first to capitalize on it, but it was taken up later by the Portuguese and the Dutch. The trade was so profitable that it inspired the search for alternate shipping routes and played a major role in world exploration from the thirteenth to the fifteenth century. In fact, the discovery of America began as a search for a better sea route to India and the Spice Islands.

The success of the spice trade depended upon maintaining a monopoly, which resulted in many small wars and deceptions. In 1512 the Portuguese seized one of the Moluccas—the island of Amboina in the Banda Sea. They built a fort there and carried on a successful trade for spices until 1605, when the fort was captured by the Dutch, who made it a base of operations from which they could control the rest of the Moluccas. It is said that one of the ships of Portuguese navigator Ferdinand Magellan, after its circumnavigation of the globe in 1519-1522, brought back a load of cloves, which was sold in Europe for more than the entire cost of the three-year trip. Amboina contained what was probably the largest clove population in the world, and on the nearby islands there were pepper vines, clove trees, and nutmeg trees. The Dutch invaded these islands under cover of darkness, destroying the trees and vines in order to increase the scarcity of the spices and run up the prices.

For many years this trade, conducted by the Dutch East India Company, was successful. Cloves commanded enormous prices in Europe, and to keep prices high, the traders invented a number of ingenuous stories and superstitions. For example, nutmeg trees, they said, grow only where there is the sound of surf, and cinnamon trees grow only in the center of a mysterious lake, guarded by birds as large and ferocious as mythical dragons. In order to harvest cinnamon, the birds had to be chased away at great personal peril, then the harvesters would wade into the lake, tear off a few branches of the cinnamon trees, and hurry back to safety.

Another story was told to customers in Europe. None of them had ever seen a cinnamon tree, but customers were told of certain eagles who built nests of cinnamon branches. The only way to get these branches was to lure the birds with heavy pieces of meat, which they carried back to their nests. The nests would break under the weight, scattering the branches that would then be picked up by harvesters. Because of these dangerous circumstances, the traders said, cinnamon was very rare and very dear.

Many wonderful healing powers were also assigned to cinnamon, nutmeg, and cloves. They were believed to be capable of curing a long list of human ailments, from leprosy to insomnia.

The Dutch managed to maintain extravagant prices for spices until the French broke their monopoly in the eighteenth century. At that time, the Frenchman Pierre Poivre, an administrator of Ile de France (now Mauritius), paid an official visit to Amboina and asked politely to be shown the nutmeg plantations. Obligingly, the governor drove Poivre around the island to view the trees full of ripe fruit. When he left, Poivre had both nutmeg and clove seeds hidden on his person. He carried these to Mauritius, and many trees descended from them on Mauritius and other islands in the Indian Ocean. In fact, today Zanzibar, off the east coast of Africa, is the largest producer of cloves on Earth.

Thus, these important endemic species were transplanted to similar habitats around the world and were never threatened with extinction. Too valuable to be allowed to suffer such a fate, they were cultivated with care and have formed a close partnership with mankind.

Through the mists of legends, witch doctors' magic, and old wives tales, the true story of the benefits of a wide variety of flora began to emerge, including the hidden potential of the many endemic species that have evolved in obscurity on isolated islands around the world. For example, the periwinkle, a small pink flower that grows in the south of Madagascar, is the source of a precious extract that arrests childhood leukemia. There are many other periwinkle species and nearly 10,000 flowering plants in Madagascar. Their potential has never been tested, and we can only imagine what we might find if these plants survive long enough to be carefully examined.

I believe that the Wise Men who came out of the East, bringing gifts of frankincense and myrrh to honor the Christ child, were also bringing an important message to mankind, a clue to the treasures that may still lie hidden in many fragile species that will be lost forever as the world becomes more crowded and more competitive. Soon there will be no islands left that can provide safe habitats for the most vulnerable, the least fit, species. The pink periwinkle has no thorns to protect itself but it carries a gift of life to the children of the world.

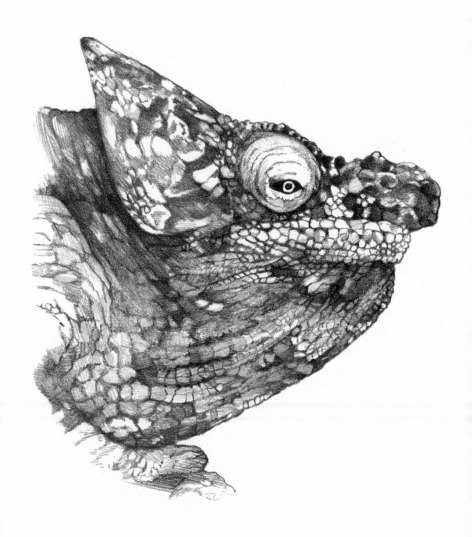

Chapter 6

MADAGASCAR

The Land Apart

Nature hath framed strange fellows in her time.

—WILLIAM SHAKESPEARE

In 1993, on a dry, windswept savannah on the northwest coast of Madagascar, a team of paleontologists discovered a treasure trove of fossils buried in a nondescript grassy hillside. This place in the Mahajanga Basin would become one of the most significant paleontological sites in the world. Expeditions of scientists working during the subsequent years found and identified the fossil remains of many species: turtles, snakes, lizards, frogs, bony fishes, sharks, rays, and crocodiles. There were also bones of several dinosaurs—carnivorous theropods and large herbivorous sauropods—and dating showed these fossils to be 75 million years old.

The technique of dating the age of the rocks in which fossils lie depends on the remarkable phenomenon of radioactivity. Each radioactive element decays (undergoes changes in its nuclear structure) at an absolutely constant rate. Therefore, by measuring the proportion of the parent atoms (those that have not undergone radioactive change) to the daughter atoms (those that have changed), the age of the rock can be determined.

One of the greatest prizes found in the Mahajanga Basin site was the skeleton of a previously unknown carnivorous dinosaur, which they named *Majungasaurus*. Probably bigger than a rhinoceros and weighing several tons, it walked on its hind legs and was a ferocious predator, equipped with large serrated teeth like the blade of a steak knife. The paleontologists were surprised to find that this fossil's jawbone had characteristics very similar to *Indosuchus*, a genus of dinosaur found in India. It also resembled (although not quite as closely) a group of dinosaurs discovered in Argentina.

The fossil discoveries confirm, in general, theories about continental drift, and they also throw new light on the timing of the movement of continents and provide new information on the way species evolved and moved around the changing planet. Geologists had believed for some time that Madagascar split off from the African mainland between 90 and 100 million years ago, at a time when dinosaurs still ruled the Earth, so it was not surprising to find dinosaur fossils in Madagascar. But the presence of a dinosaur that appears to be a near relative of some that were found in India and South America suggests that there must have been a route of migration across the land masses of the southern continents.

Two hundred million years ago, the continents of the Earth were united in one single unit, named Pangaea. Slowly but irresistibly, the large continent was broken apart by tectonic rifting, so that by 160 million years ago, the northern land masses had separated from the southern ones, which were still united into one giant continent, Gondwanaland. South America, Africa, Antarctica, Australia, India, and Madagascar were all joined together in a contiguous formation. As a result, there would have been a land-based path for the movement of animals from Argentina to Madagascar or India. For example, an ancestor of *Majungasaurus* could have passed from Argentina to South Africa, Antarctica, India, or Madagascar without having to cross an ocean barrier.

More massive rifting then took place, tearing Madagascar, India, Australia, and Antarctica from eastern Africa, creating a separate land unit by about 150 million years ago. Seventy-five or 80 million years later (exactly when is still a matter of debate), Madagascar and India separated from Antarctica and Australia, and, finally, India broke away from Madagascar and traveled northward until it collided with East Asia. Madagascar, like a crumb left over after a cookie was broken into many pieces, remained in virtual isolation for at least 70 million years. It is one of the oldest islands on Earth.

But the most significant discoveries unearthed at the Madagascar site were the bones of four very primitive birds. The bones are small and thin-walled with several distinctive avian features as well as several associated with dinosaurs. One of the four fossils—more than half a skeleton—is truly the gem of the treasure trove. Although other fossils have been found that appear to be missing links between dinosaurs and birds, this one is the most complete. It is about the size of a hawk and has a long tail and a large retractable, sicklelike claw on the second toe of its hind foot. This claw is a smaller but identical replica of the killer claws wielded by small carnivorous theropods like *Velociraptor*. These striking similarities lead to the hypothesis that this bird was descended from a theropod dinosaur.

As a result of these significant discoveries, a curtain has been lifted on the history of the Earth, as well as on Madagascar's past. For 10 to 15 million years these dinosaurs, primitive birds, various amphibians, snakes, crocodiles, and fish shared this isolated island habitat until they were subjected to one of the most catastrophic events in all geologic history. Sixty-five million years ago the impact of a large asteroid was the probable cause of a massive die-off of many life forms around the Earth. Most—perhaps all— of the dinosaurs became extinct, and many large marine reptiles, like the ancient crocodile, and most marine plankton and marine organisms, like the ammonites, also died out.

The extinction of large, dominant, and, in many cases, carnivorous species offered an opportunity for the smaller life forms, like the mammals, to multiply and evolve into successful species in many habitats around the world. These early mammals were small rodentlike creatures that are known only from isolated fossil teeth and jaws—not one complete skull has been found. In fact, two mammal teeth similar to those from the ancestors of lemurs were found at the Madagascar site. The earliest fossils of lemurlike ancestors come from Europe and North America, dating back about 60 million years. However, evidence of their presence in Africa at that time is very controversial. Even if it can be established that lemurlike ancestors inhabited Africa, the question remains how did they manage to bridge the ocean gap that lies between the island and the mainland?

Madagascar is not far from Africa, about 240 miles, but the channel is about a mile deep in most places. Moreover, several small islands, the Comoros, which might have served as stepping stones, did not erupt from the seafloor until 7 million years ago. It has been suggested, therefore, that a few lemurs might have rafted across the ocean gap, hitching a ride on floating piles of vegetation. Although the chances of this happening seem very small, because the prevailing trade winds are in the opposite direction, we must remember that very long periods of geologic time were involved. Maybe once in 50 million years a small group of lemurs did make this trip. The event would have been most likely if they had been very small— about the size of the mouse lemurs that still exist today. Therefore, it is mysterious that the most ancient remains of lemur species (about 5000 to 26,000 years old) found on Madagascar are extremely large, some as big as gorillas. These giant lemurs are now extinct, but 32 other species are still living on the island.[7]

Madagascar as we know it today harbors the highest percentage of endemic species found anywhere on Earth. Eighty-five percent of the flora and fauna there evolved on the island. The French naturalist Philibert Commerson visited the island in 1771 and called Madagascar "the naturalists' promised land. Nature seems to have retreated there into a private sanctuary, where she could work on different models from any she has used elsewhere. There, you meet bizarre and marvelous forms at every step."

My days in Madagascar demonstrated the truth of this description, not only of the fascinating flora and fauna but the character of the human population, too. I traveled by plane to Fort-Dauphin, a port town on the south coast of the island. From the air I could see bare, eroded mountainsides and rivers with wide, meandering flood plains stained deep brick red, the color of the iron-rich soil washed down from the mountains. The town was rust colored, too, and black. The buildings were made of brick or corrugated iron. Many of them were tiny shacks no larger than a doghouse. Poor, ragged, and very black children filled the streets. Small girls, some as young as eight or nine, carried babies on their backs. The adults were clean-looking despite the impoverished conditions in which they lived. Many of them wore white shawls draped across their shoulders like togas. A few were dressed in varied and beautiful costumes.

Because Madagascar is a very large island, it encompasses several climate zones. The southern part is extremely dry. Rain falls there only in December, January, and February, totaling about 12 inches. But even this rain does not penetrate far into the soil, which is baked dry as though it had been fired into bricks. There is simply not enough vegetation in the bare hills to hold and distribute the moisture.

Zebu cattle and pigs wander in the streets. The zebu are humpbacked cattle well adapted to a hot, dry climate.

They are especially valuable to the Malagasy (the people of Madagascar). They serve as beasts of burden to pull oxcarts, and they trample the earth in the rice paddies, churning it up to make it suitable for planting. Most importantly, they are used as sacrifices in ancestor worship, a dominant part of the native religion.

Although most of the people are nominally Christian, they combine this with their traditional beliefs. They believe in one God and the continuation of the spirit after death. The souls of the ancestors rule the family life. If illness or other misfortune occurs, it is because an ancestor is angry and must be propitiated by a sacrifice. When a member of the family dies, zebus are sacrificed at the funeral. The more important the person, the more animals are slaughtered. Their skulls and horns are hung in a memorial cemetery above little wood carvings depicting the episode that caused that person's death, perhaps by drowning or being eaten by a crocodile. The body itself, however, is interred somewhere else and the bones are dug up every three years or so. They are paraded through the village, wrapped in a new shroud, and are treated with great ceremony before they are buried again.

The Malagasy have many unusual customs and taboos known as *fadies*. If a person is caught perpetrating a fady he will probably be punished by death.

The people are distinctive in many ways. Their skin is the blackest black, like bitter chocolate, but their features are not African; they more closely resemble those of the Polynesians or Malaysians. In fact, the Malagasy are believed to have arrived in progressive migrations from areas farther east. Sailing on the southeasterly trade winds, their fragile canoes carried them thousands of miles, like seeds borne on the prevailing winds. In Madagascar they discovered an island suited to their way of life, with a more favorable climate and perhaps a much larger land area than the places from which they had come. Although there must have been some interchange of people from the mainland of Africa, which is so much closer, this inheritance did not contribute as much to the Malagasy people. The Malagasy are yet another example that the historic distribution of life around the world depends on which way the winds blow.

The environment and weather conditions of the different areas of Madagascar are similarly influenced by the direction and nature of the wind flow. The southern portion of the island lies at 25 degrees south latitude. This is a region of high pressure, where downward air movement is dominant and most of the water is condensed out when the sinking air masses are warmed by compression. In Madagascar, as in the other desert regions of the planet, the vegetation has adapted to these dry conditions by evolving ways of storing water and reducing the amount of evaporation through the shape and area of their surfaces exposed to the atmosphere. As a result, many cactuslike plants have evolved.

Northwest of Fort-Dauphin is an area where striking examples of this type of flora dominate the landscape. This region, known as the spiny forest, is densely occupied by tall green spikes, armored with a million thorns. Although they look very similar to cactus plants in the Americas, they are quite different genetically. Many of them seemed to be built on a very primitive principle: With straight trunks, very few branches, and no leaves, they are ideally designed to draw the moisture up from the ground. These plants are not unlike those very early vascular flora that inhabited the Earth 400 million years ago. As I stood there, surrounded by these weird shapes, I felt as though I had been suddenly transported back to Paleozoic times. I half expected to see a dragonfly with a 3-foot wingspread flying overhead or perhaps an amphibian like *Lystrosaurus* scuttling along beneath this green mass of thorns.

Beyond the spiny forest the strange vegetation suddenly gives way to sisal plantations. Field after field, very neatly and systematically laid out, stretched as far as the eye could see. These are all owned by one French family that also owns the two principal hotels in Fort-Dauphin and the Berenty Lodge and Reserve.

The Berenty Reserve is a beautiful wooded area along the Mandrare River. Tall tamarind trees provide the highest canopy, casting a light shade, and there is no ground cover except in places where the rubber vines have completely covered the bushes and smaller trees. There are fig trees with their strange spaghettilike structure and lianas making loops and figure eights in the shadows. Many epiphytes decorate the branches of the taller trees. Hundreds of ring-tailed lemurs play in and among these branches, down the long, shaded paths. They follow us because we have brought bananas—their favorite food—and they climb right up into our arms or perch on our backs, showing no fear at all. With their big, golden eyes in little pointed faces, their startlingly human-like hands, and their magnificent long, striped tails, these charming creatures are very friendly and gentle. Their sharp teeth and claws are carefully managed, so even when they are scrambling up to get a banana, they do not scratch or bite. Many of the females carry a baby, either slung beneath (if the baby is very young) or riding on the back, until they are about six months old. (I am reminded of the little girls I saw in Fort-Dauphin with babies slung on their backs.) The lemur mothers carrying their babies leap and swing from limb to limb with ease and grace. It is a remarkable sight to look back down a tree-shaded path and see a whole troop of lemurs, striped tails raised high, following in our footsteps.*

---

*The feeding of bananas to lemurs is no longer allowed in the Berenty Reserve, and I am glad that the health of the lemurs is given first priority, but this rule does mean that visitors will not have the close contact with the ring-tailed lemurs that I enjoyed.

Looking into the soft amber eyes of these little creatures, one senses some recognition of a common relationship as one might see in the eyes of a long-lost cousin suddenly encountered much later down the long road of life. But it is the lemur's hands that are most reminiscent of our own. With their four fingers and opposed thumb, these small hands are amazingly human. In fact, biologists have been able to trace the relationship between lemurs and *Homo sapiens*. Lemurs are descended from the common ancestor of the line that led to monkeys, apes, and humans.

Wherever monkeys and lemurs have existed together, the monkeys proved to be the most fit in the Darwinian sense. They were more aggressive and more pugnacious. Finally, they supplanted the lemur populations. But monkeys never managed to colonize Madagascar. Protected from this competition, lemurs thrived on the island, and eventually radiated into many different species.

Several of these species are present with the ringtails in the Berenty Reserve. There are many sifaka, which are larger than the ringtails and have soft golden fur like cuddly children's toys. Their tails are long and thin but not as decorative as those of the ringtails. Because sifaka do not like bananas, they did not come close to us but remained in the branches, leaping agilely from tree to tree.

One particularly interesting species is the black lemur. There is a colony of them on a tiny island off the coast of Madagascar, Nosy Komba, the "island of lemurs." Black lemurs live there just on the edge of a native village where they exist in harmony with human beings. The people in this community revere the lemurs; it is a fady to kill one, and, therefore, the lemurs are friendly and trusting. Only the males of this species are really black. The females are slightly larger, and they are a handsome reddish-brown with cream-colored ruffs;

they are the dominant members of this species. Although most of the females that I saw were carrying babies, this burden did not impede their leaps and jumps from tree to tree. These lemurs, as I said, also love bananas, and on Nosy Komba feeding this favorite food is allowed. When I was there, everyone who held a banana had a lemur in his arms.

The largest and most reclusive of all the lemur species lives in the Perinet Reserve on the eastern shore of Madagascar. This is a small remnant of the rain forest that once covered the whole eastern seaboard. It provides a tiny shelter for the Indri lemurs. Deep in the jungle these beautiful, shy creatures enjoy their domain high in the treetops.

The Indri cannot be kept in captivity. They flee from humans, do not tolerate any disturbance of their habitat, and eat the leaves of about 60 different kinds of trees—a diet difficult to provide in a zoo. The Indri are the ultimate aerial acrobats, leaping from the crown of one tree to another, never touching the ground. With a powerful thrust from their legs, they stretch out their long bodies and sail through space, ricocheting from tree to tree, until they finally settle among leaves so dense that they become almost invisible. Using binoculars, I could see that they are about 3 feet long with very short bobbed tails. Only their faces and ears are dark while most of their bodies are buff-colored.

At least once every 24 hours, usually in the middle of the morning, the Indri sing. The sound is eerie and beautiful; it makes the forest spaces ring as though struck by a great bell, as one after another lemur takes up the song. The meaning of the activity has never been decided by scientists. It may be territorial, or it may pass along important information about the presence of danger. But then it may be just a way of maintaining contact with others, relieving the loneliness of their solitary journeys through space.

Some biologists have suggested that the song is a way of judging the size of the community—a sort of census-taking. The population density may be a reason for concern, because the Indri reproduce very slowly—on the average a single birth occurs every three years—so the population is small, with all the jeopardy that rareness entails.

The diversity of the lemur species in Madagascar is extraordinary, considering that they have all evolved on this one island. But perhaps this fact is not so surprising. Madagascar provided an ideal situation for the evolution of many species. It was isolated for a very long period of time, and it did not have important predators that could have decimated the populations of gentler species. There were no lions, tigers, or leopards—not even any poisonous snakes. Furthermore, the island was large enough to support viable populations of individual species. And many diverse habitats created diverse evolutionary pressures. When mutations occurred, those few that improved the adaptation to the environment of that particular population were passed down in greater numbers and gradually a new species evolved.

The phenomenon of many closely related but distinct species is manifested in myriad ways on Madagascar. Trees, for example, are present in an amazing array. Six species of baobab tree are found only in Madagascar, while Africa has just one. The largest, called the "mother of the forest," have huge, cream-colored trunks like temple columns. Their smooth trunks are tall, shining features in the forests. Water is stored in their trunks, so these trees can survive sparse rainfall and provide a source of water for animals and humans during periods of drought. One species of baobab tree is especially valuable because its wood is resistant to termite invasion—a trait that should have great economic value. But many baobab trees are falling to the axes of the peasant farmers, and it may be too late to take advantage of their unusual characteristics. Like the Indri, the baobab trees have a very slow reproduction rate. Many do not bear fruit every year and only once in every five or ten years do the seedlings of a given species survive for more than a few months.

Many different types of palm trees also grow in Madagascar. The three-cornered palm, for example, is very rare and confined to one small area of the island. The traveler's palm also evolved here, but because of its unusual and beautiful shape it has been exported widely to other countries.

Another fauna that has radiated into a spectacular number of species on the island is the chameleon. Survivors from very ancient times, these creatures change color according to their mood and their surroundings. They are present in astonishing variety in Madagascar. One type has a tongue that can shoot out a body-length away to capture an unsuspecting insect. Another is frequently colored a brilliant green and has moody green eyes and a tail that is curled in a counterclockwise helix. Half the chameleon species in the whole world are present here.

Flowers grow on the island in numerous varieties as well. There are said to be 10,000 flowering species, including nearly 1000 unique varieties of orchids. Among this great storehouse of evolutionary wealth are several that hold unusual potential for mankind. As we have already noted, one of the periwinkle species produces a chemical extract that arrests childhood leukemia. There may be many other treasures that have never been evaluated.

Several plants have been imported and grown very successfully in Madagascar, because their harvesting depends upon vast amounts of cheap hand labor. The elan-elan tree, which evolved in the Philippines, is now raised in extensive orchards on Madagascar. The trees are pollarded (cut back at the top), and thus are forced to branch out close to the ground, so the blossoms can be reached more conveniently. The pale yellow blossoms are very fragrant, and essence is extracted from them to use in many perfumes, although 100 kilos of blossoms produce only 2 liters of essence!

The vanilla plant, which belongs to the orchid family, is a rather undistinguished-looking vine, imported from Mexico. In Madagascar it lacks its symbiotic partner (a special variety of fungus), and the blossoms must be pollinated by hand. Despite this difficulty, Madagascar is now the world's principal source of natural vanilla.

Throughout the island, especially in the nature reserves, there are beautiful endemic birds: the giant coua, with an iridescent blue cap on its head; the crested coua, which looks like a very large bluejay; the paradise flycatcher, which nests in the tamarind trees; the drongo, a striking bird that is dark-colored with a forked tail and a saucy little crest on the top of its head.

Until quite recently, one particularly bizarre-looking bird inhabited Madagascar. Like a mythical character in a children's book, it stood about 10 feet tall and weighed perhaps 1000 pounds. It was undistinguished in shape, like a vastly overgrown chicken. It could not fly, its wings tiny and ineffectual, but it had powerful legs. Aptly named the elephant bird, its fossil remains, as well as an enormous egg with an estimated volume capacity of about 2 gallons, have been discovered. These are now on view in a museum in the capital city of Antananarivo. Even this unlikely creature had a number of closely related but distinct species living on the island—a handful of slightly different, giant flightless birds.

Birds like this (known as ratites) have existed in many places around the world: the dodos of Mauritius, the moas of New Zealand, the ostriches of Africa, the rheas of South America, the emus of Australia, and the cassowaries of New Guinea. They share certain biological traits. They are unusually large in size; they have strong legs for walking, running, or kicking; they have small wings that are not strong enough to lift their heavy bodies into the air; and their breastbones lack a keel where wing muscles are normally attached. Is it possible that these birds are all descendants of a primitive ancestral form that radiated into a number of different species, adapting to diverse environmental conditions?

Alfred Wallace, who was the codiscoverer with Charles Darwin of the theory of natural selection, surmised that this might be true. In his book *The Geographical Distribution of Animals*, published in 1876, Wallace suggested that ratites descended from "a very ancient type of bird, developed at the time when the more specialized carnivorous mammalia had not come into existence, and preserved only in those areas which were long free from the incursions of such dangerous enemies." But, for Wallace, there seemed to be no ready explanation of the way this primitive ancestor could have found its way to South America, Africa, Australia, and Madagascar. Now, of course, the theory of continental drift offers a possible solution to this problem. Perhaps like the ancestor of the dinosaur *Majungasaurus*, the ancestor of the giant flightless birds was widely distributed in Gondwanaland. As this continent broke up into several separate land masses, these birds managed to survive in a few isolated situations where the competition with predators was not strong enough to decimate their populations.

This is an interesting theory, although it has not been completely accepted. New theories in science must always be tested through a long period of questioning and debate. As Wallace himself said, "Truth is born into this world with pangs and tribulation, and every fresh truth is received unwillingly. To expect the world to receive a new truth, or even an old truth, without challenging it, is to look for one of these miracles which do not occur."

Whatever the ancestry of flightless birds, when the island habitats like Madagascar and Australia were invaded by mankind, these birds, which were easy to capture and provided an abundant source of food, did not long survive.

Madagascar itself may suffer a similar fate. As one drives across the countryside, the gaping open wounds caused by human molestation are apparent everywhere. Mountainous territories that are totally denuded of trees bear deep red scars where eroded hillsides have slumped into the valleys. Rains no longer slake the thirst of these hillsides. The water rushes over the hard-baked ground in torrents, carrying more topsoil into the rivers and eventually into the ocean, which is blood-red and murky where rivers flow into the sea.

The evidence of fires is apparent everywhere, because the Malagasy use the slash-and-burn method of farming. Burned-over land provides a "green bite" when it comes to life again, providing pasture for pigs and goats and for the ubiquitous zebu cattle. When the land becomes exhausted, the Malagasy simply pack up and move on.

Like poor peasant farmers all over the world, the Malagasy cannot understand why these methods are discouraged by environmental experts from developed lands. They see the issue as a battle for survival between the forest and the people. The subsistence of their children is more important than the saving of rare endemic species. A baobab tree or a child—which is more valuable? Stated in these terms, it is hard to convince the people that by destroying their environment they are committing suicide.

The economics of the problem are quite simple and easy to understand. For a minimum survival diet, each person must eat a bowl of rice three times a day. If this amount of rice were purchased, it would cost the average Malagasy family three-fourths of its annual income. Obviously, this is too great an expense for the native population. The alternative is for each family to raise its own rice.

As we drove through the countryside, I saw that every little valley was filled with rice paddies, and the seedlings were the most vivid velvety green that can be imagined. Here and there a small field of wheat was set in among the rice fields. Rice farming is an especially labor-intensive occupation, with many women planting and tending the young seedlings by hand.

As the already large population grows, more and more space is needed for planting crops. Trees must be cut down; the land must be cleared; and after the rice is harvested, it must be cooked. Diesel fuel or kerosene is too expensive for the natives, so more trees are sacrificed to provide charcoal. Thus, the destruction of the forests becomes inevitable. It would appear to be the story of Easter Island all over again.

With the remarkable achievements of modern technology, it should be possible to meet the comparatively simple challenge of helping the people of Madagascar prepare the rice to feed their families without destroying the forests that are essential to preserving their environment.

Abundant energy is available—both free and renewable. The island is flooded with sunshine; the trade winds blow with predictable regularity; and the mountain streams bring surges of water power down the steep mountainsides. Many innovative ways of harnessing these sources of power have been invented, calling for only a modest investment in equipment. For example, a very inexpensive little device can be made with a few ounces of aluminum to focus the rays of the sun and produce heat intense enough to boil water and cook rice. The gift of such a tool to each Malagasy family could save the precious remaining stands of forest.

If this kind of technology were used to solve the problem in Madagascar, it could be used to good advantage in many other undeveloped countries around the world as well. Some of the most desperate situations are in regions where the sun shines with unrelenting strength, for example, India and the valley of the Amazon. Innovation, intelligence, and dollars wisely spent could save our planet as a dwelling place for mankind.

On our last evening in Madagascar, we drove to a mountaintop overlooking the western sea on the little island of Nosy Be. It was a gentle evening with just a few cirrus clouds in the sky, which turned a deep rose color as the sun set over the Mozambique Channel and farther (beyond our sight), to the Comoro Islands and the coast of Africa. The land where I stood was fractured with dark sharp edges of lava flow—an old volcanic crater. Below and to the south, there were three crater lakes, their waters pink and tranquil in the sunset glow. Just beyond the horizon is the Mahajanga Basin where ancient history is still embedded in the sand, including the bones of huge reptiles that once roamed these shores—the heavy sauropods, the carnivorous theropods with their killer claws, the giant crocodiles, and a whole assembly of little creatures that passed their lives in the shadows of these monsters. Sunset after sunset passed for 90 million years. Time has worked its magic and everything has changed. The craters, where boiling lava once steamed and flowed, are quiet now and cool with surfaces barely riffled by the wind.

As I watched, a flock of egrets rose from a grove of trees between the crater lakes. They flew high above the sea, their graceful white wings reflecting the pink of the sunset. Where once the *Velociraptor* trod on heavy feet, time and evolution have wrought this miracle. If all the millions of years could be condensed in one moment of revelation, the flight of the egrets would seem like a rainbow appearing in a darkened sky—a covenant with Nature, a symbol of hope.

Chapter 7

MAURITIUS AND THE SEYCHELLES

Home of the Dodo

Some unsuspected isle in far-off seas.

—ROBERT BROWNING

Five hundred miles east of Madagascar, in the tropical, exotic heart of the Indian Ocean, lies the little island of Mauritius, as young and promising as Madagascar is old and scarred. Its varied landscape still bears the afterbirth of its violent nativity about 7 million years ago. Seen from the air, this residue presents an unusual aspect. In the fertile plains between the mountain ranges, each field contains a dark pyramid of lava rocks piled about 20 feet high. And around these black mounds, acres of sugarcane wrap a vivid green perimeter.

These volcanic "bombs" have been laboriously collected from the surface of the fields in order to farm the land more efficiently, and the arduous work has paid dividends in successful farming. The agricultural produce has been sufficient to maintain an active export trade as well as to feed a large human population. Mauritius is one of the most densely populated regions of the world. With 1390 people per square mile, it is almost as crowded as Bangladesh.

Recall that Easter Island was also strewn with volcanic debris, but when the ruling minority demanded that these rocks be gathered up and thrown into the sea, the laborers revolted, and the war that followed resulted in the ultimate demise of the civilization on Easter Island.

The human story of Mauritius began just a few centuries ago. The island was unknown to Europeans and perhaps to all other peoples until the year 1505, when it was discovered by a Portuguese navigator. Mauritius lies on what were the early trade routes between East Asia and Europe by way of the Cape of Good Hope. It was a convenient stopping place for the sailing ships that plied these routes—a place to reprovision and restore supplies of fresh water. Not only did the sailors find the island uninhabited, but there were no traces of previous occupation. During the rest of the sixteenth century, it remained a Portuguese possession. However, no settlements were made there, and in 1598 the Dutch took possession, only to abandon it in 1710. A few years later, agents of the French East India Company took control, and the island was finally developed. The French cleared most of the dense forests that occupied the fertile lands, using slaves imported from Africa and Madagascar, and then set up sugarcane plantations, which became the main occupation of the island.

In 1810 the English conquered Mauritius, and in 1835 they emancipated the slaves. But because the planting and harvesting of sugarcane is labor intensive, Indian laborers were soon imported to work in the fields. These workers were paid so little (about a dollar a month) that their situation was not very different from slavery.

During the nineteenth and twentieth centuries, this society, consisting of many diverse factions, survived a number of disasters, including plagues of cholera, malaria, and smallpox, as well as several very violent typhoons. (Tropical cyclones in the western Pacific or the Indian Ocean are called typhoons, while in the Atlantic or the eastern Pacific they are called hurricanes.)

Because typhoons frequently strike Mauritius, the buildings there have been designed to offer maximum protection from wind. They are low and flat-roofed and constructed of concrete blocks.

Despite all these hazards, the society that emerged on this island was unusually successful. A peaceful relationship still exists between the various racial and religious factions. There is a sense of laissez-faire, a feeling of live and let live, and harmonious separatism is the unwritten law of the land. Creoles (mixed French and African ancestry) are found in a variety of occupations, the Chinese are merchants, the Hindus manage the politics, and the whites (mostly of French origin) run almost all of the big sugar plantations. Amazingly, these factions share amicably the island's facilities—the beaches, restaurants, schools, and public buildings. They all help to celebrate the special holidays of the other religious groups. (We saw a similar tolerance on several of the Indonesian islands, such as Bali and Lombok.) For example, Muslims celebrate Dewali, a Hindu holiday; Hindus celebrate 'Id al-Fitr, a Muslim holiday; and everybody celebrates Christmas. Typical of the Mauritian spirit of religious tolerance, Hindus, Christians, and Buddhists line the streets to watch the parade of Muslims at the yearly holy day that marks the end of the ten-day Yamse festival. In this parade, mortification of the flesh is endured in public as devout Muslims walk slowly in deep trances festooned with arrays of long needles and hooks driven into their bodies.

Although the races rarely socialize with each other, they work amicably together and their economy is booming. Twenty-five years ago it suffered from chronic unemployment and the population was growing very rapidly. The society seemed destined to repeat the disasters of Easter Island, but then remarkable changes took place. Unemployment has dropped from 20 percent to about 3 percent; per capita income has doubled; and the economy continues to grow. One of the important factors that brought about these changes was an improvement in the condition of women. Emancipated from perpetual childbearing and virtual slavery at home, they have taken advantage of the educational opportunities and entered the workforce. The government has undertaken a family-planning campaign. Today, the population growth rate is about 1 percent, or one-third the annual increase three decades ago.

Although these developments are remarkably positive from the human standpoint, the booming economy and large population have put an unusual stress on the island's natural resources and wildlife. The advent of human beings, who brought their dogs, cats, chickens, and goats with them, completely changed the ecology of the island. Suddenly, predators were an important part of the ecosystem.

This drastic alteration took its toll on the fauna that had evolved in this isolated land. Like Hawaii and Madagascar, many of the species were endemic—a treasure chest of the strange and wonderful—but one after another, certain of the treasures have become extinct. One example is the giant tortoise, similar to those found on the Galápagos Islands. When the Dutch and French settled Mauritius, there were many of these enormous slow-moving creatures. They were easily caught and provided a readily available source of food. Thousands were butchered, and the flesh was salted or rendered for fat. Giant tortoises also could be stored alive, because their endurance allowed the animals to linger for months in the hold of a ship. To prevent them from wandering, they were turned upside down. Thus, these vulnerable and relatively gentle beasts were ideal for provisioning a ship for a long voyage. Within a short time they became extinct in the wild, and today they are preserved only in protected situations.

Another victim of the arrival of mankind was the dodo, which was present on Mauritius as well as on Réunion and Rodriguez (two small islands near Mauritius). The first Portuguese and Dutch settlers found the islands inhabited by a large, ungainly creature—a flightless bird like a pigeon but 20 times larger. This bird was fat, its movements were clumsy, and its wings were extremely small. It was unable to fly, and, therefore, it provided an easy and abundant source of food for the settlers.

How did such an improbable creature come to be? According to one popular theory, the dodo was descended from a ground-feeding variety of pigeon that had evolved in isolation after flying to the islands, and as they evolved they lost the power of flight.

We might consider the loss of flight a biological disadvantage and wonder how it could have occurred by natural selection, but at the time this happened in Mauritius, flight may not have conferred an advantage. Before mankind arrived, this island provided an ecological niche where there were no mammals (except one bat) and no predators to threaten a bird that spent its whole life on the ground and laid its large single eggs there. Wings were not needed to escape predators and might have been an added burden or even a nuisance. For example its wings could have caught on low-growing vegetation.

Or perhaps this bird was simply too stupid or too lazy to take advantage of the gift of flight. This possibility is supported by the impressions of those who observed the bird in its final state. They called it "dodo" after the Portuguese *duodo*, meaning simple or foolish. There was general agreement that there was something ludicrous about the dodo.

Another popular theory is that the ancestors of the dodo, having settled on the island, put on weight because food was always readily available. As they got heavier, their small wings were less capable of lifting them into the air. Gradually, they lost the power of flight, and this loss made them easy prey for the settlers and crews of sailing ships that stopped in Mauritius to reprovision. By 1625 the dodo was completely extinct.

The explanation of the origin of the dodo is actually based on informed guesswork; there is no fossil evidence that proves it correct. There is another theory worth mentioning, however. Perhaps, as Alfred Wallace suggested of the ratites, this bird never did possess the power of flight. Like the fossil bird discovered in Madagascar, the evolutionary changes that transformed a reptile like the theropod dinosaur into a primitive bird had not yet produced a bird able to fly, although they had achieved other avian characteristics. Perhaps the dodo was descended from a transitional organism of this kind. It did not lose the power of flight—it never had it.

The trouble with this hypothesis is that it is difficult to imagine how such a primitive "bird" reached Mauritius (and also Rodriguez and Réunion), which had erupted in volcanic action long after the dinosaurs had become extinct. Maybe they were rafted to the islands, blown in by a typhoon, which is a very common phenomenon in that part of the world.

This possibility was regarded with some skepticism until very recently. While there had been records of single, unusually small animals drifting on rafts of vegetation, the chances that this method of transportation could have resulted in the introduction of a large species like the dodo on an isolated habitat seemed unlikely.

In 1995, however, a surprising event was witnessed. Fifteen large iguanas arrived on the little island of Anguilla in the Caribbean, having floated on a raft of trees 200 miles from the island of Guadeloupe. These fearsome-looking animals, similar to small dinosaurs, measured up to 4 feet long. They were still alive, although weak and dehydrated. Scientists believe that one of two powerful hurricanes that struck Guadeloupe in September 1995 could have uprooted large trees, which fell into the sea, carrying with them iguanas that had been clinging to their branches. The track of the storm brought them almost a month later to Anguilla.

After their arrival, the animals recovered and gradually established themselves. In March 1998 one iguana was found to be pregnant, so this chance phenomenon represents a successful colonization of a new species, and it shows that a species need not be especially small to make a sea voyage successfully. This is an important discovery that has revolutionized the theories about the transport of species to various island habitats. It answers a question that we asked in Chapter 6: How did lemurs

get to Madagascar from Africa, especially when the earliest ones are believed to have been as large as gorillas?

In every environmental niche there is an interdependence between the various species that occupy it. The extinction of one species affects the environment of the others, and in some cases the effect may be very unfavorable. An interesting theory has been advanced about the disappearance of the dodo and the survival rate of the endemic fruit-bearing *Calvaria major*. It was once common in the upland Mauritius forests, but by 1973 very few could be found. The seeds of this tree are surrounded by very thick-walled pits inside a sweet fruit. The ecologist Stanley Temple theorized that this trait evolved as protection against the ravenous appetite of the dodo bird and the powerful grinding power of its digestive system. As the pit passed through the bird's gizzard, it was subjected to enormous abrading pressure. A more fragile pit would have been destroyed along with the seed in the dodo's gizzard. In response to this challenge, the Calvaria developed tougher and tougher pits, so at least a small proportion of them could survive passage through the dodo's digestive system. With the pits only slightly abraded, the seeds could push their way out and germinate. Thus, the tree was dispersed throughout the island's habitats. But when the dodo became extinct 300 years ago, the Calvaria tree also seemed to disappear from the island. Had the Calvaria's seeds become prisoners wrapped in their own pits?

This interesting theory, however, has been challenged by several experts who have identified a number of young Calvaria trees in the forests, and they suggest that other birds, such as a very large endemic species of parrot, could provide the service previously rendered by the dodo. Nevertheless, it is true that the Calvaria tree is becoming exceedingly rare. And the loss of the cooperation of the dodo, although perhaps not exclusively responsible, may have made a difference in the chances that these seeds would germinate.

At any rate, the concept of a mutual relationship between the two species is a significant one in evolutionary theory. A surprising interdependence has been shown in many cases. For example, the diet of Indri lemurs must include the leaves of many endemic trees in Madagascar, and the panda depends upon the foliage of the bamboo tree. Obligatory mutual dependence is a disadvantage for long-term survival and is more often found on islands where competition is at a minimum. In more challenging environments, species must develop the ability to adapt to changes in the ecosystem in order to survive.

The environment on Mauritius has undergone drastic and very rapid changes. Although quite a few of the endemic species that evolved there are now extinct, some still remain and these are severely threatened. Originally, Mauritius was enfolded in forests that provided a protective green blanket that soaked up the sunshine, moderated the winds, and held and distributed the rainfall. There were dense hardwood stands of ebony, teak, and mahogany. Many different species of palm and banyan trees decorated the coastlines and the edges of the forests.

Homo sapiens and trees have had a special relationship for perhaps 3 million years and, for the ancestors of man, perhaps several million more. As we have seen with the lemurs today, trees provide a home, food, shelter, and some protection from the elements and predators. When threatened, they retreat to trees, where they are concealed by the dense foliage. The trees create shadows and cool, shady places, where the intense rays of tropical sunlight cannot penetrate.

It has been suggested that trees were an important factor in the evolution of primates. By climbing and leaping from branch to branch, they acquired agility and developed strong arm and leg muscles that could act independently or in unison. Challenged by the need to grasp the branches and pluck the fruit, their hands evolved the favorable shape of four flexible fingers with an opposing thumb. Stereoscopic vision is also important for arboreal life, because it is an advantage in judging distances. The refinement of these characteristics resulted from their life in the forests. When they came down from the trees, the early primates were equipped to walk upright, to pick up and manipulate many objects in their new environment, and to gather and shape stones and to fashion tools.

But was this relationship mutually beneficial? What have human beings done to promote the vitality and adaptation of the forest? Sadly, not very much. To some extent, people have served as a transportation system for the seeds of the forest. They eat the fruit and throw away the seeds. They have planted a few specimen trees in parks and yards, but these have not been important contributions. At the same time, people have played a major role in the destruction of forests. In Mauritius less than 1 percent of the native forests remain.

Wherever primitive mankind settled, trees were destroyed. The principal reason for the taking of trees was to build homes (improving on the advantages that the forest originally conferred by providing shelter and protection from predators and the elements). Later, of course, the wood was used to create additional comforts and conveniences—boats, wagons, and tools. The land that the forest had occupied was used to grow food crops, and wood was used to cook the food. Finally, the trees were harvested to sell, the money providing other advantages.

We all know what the result has been. The forests of the world have steadily diminished. You might say that they are an endangered species. Imagine for a moment what the world would be like without trees. It would be a harsh, forbidding place, like the surface of Mars or the moon, where the winds would blow with undiminished strength, where the surface areas would be roasted in the sun, and where clouds of dust and sandstorms would circle the globe, abrading everything in their paths.

The varied and beautiful outlines of the planet as it exists today would be stripped of the soft shapes against the sky, and the special beauty that returns to delight us every spring, when the leaves and blossoms begin to come out on the deciduous trees. The shape of each bud and opening leaf is seen separately, as though framed in three dimensions, and each tree is an artistic construction in space. How much would we lose if trees became a rare species and forests became extinct?

The loss of forest cover has affected many of the endemic birds that make their home in Mauritius. Ornithologists have concentrated their rescue efforts on the two species that seemed to be the closest to extinction—the Mauritian kestrel and the pink pigeon.

The kestrel is a member of the falcon family but looks more like a hawk. It is dark brown and quite large, with short, stubby wings adapted for brief, quick flights between trees. Because of habitat loss, poisoning by DDT, and egg predation by monkeys, rats, and mongooses, which were imported from Asia to control rats, the population of kestrels steadily declined during the 1970s, to the point where only two breeding pairs remained.

In that decade, the International Council for Bird Preservation (ICBP) launched a conservation effort to prevent the extinction of this species. During 1978, in the wake of a disastrous typhoon, only one pair reared young in the wild. The rescue attempt seemed doomed to failure, and the ICBP withdrew its support, but the cause was taken up by the Mauritian Wildlife Fund. Several young English and American ecologists accepted the challenge and used some imaginative techniques to improve the chances of success. The locations of kestrel nests were identified on ledges high in the cliffs. In a perilous maneuver, the ecologists scaled the sheer granite mountainsides and reached the nesting sites. There they removed one or two eggs and left an egg laid by a European kestrel. In this way the female was not too discouraged and continued to lay more eggs. The original eggs were then carefully carried back to the research station, where an aviary had been built with equipment for hatching them. Day after day the stolen eggs were brought back. The female frequently laid as many as eight instead of the usual three. The nestlings in the aviary were raised with care and fed a diet enriched by raw meat—minced quail was a favored specialty. The hand-raised kestrels tended to remain in the vicinity of the station when released. They could be trained to return for a special treat offered by the staff.

I witnessed an amusing little stunt invented by Carl Jones, the leader of the team. Approaching a kestrel's favorite roosting site carrying several dead white mice, he would whistle to the kestrel and toss a mouse into the air. The bird would dive off its perch and swoop down to catch the mouse in midair above Jones's head.

Over the next few years the rescue team pursued their goal with great dedication and were eventually successful in saving the Mauritian kestrel from what had appeared to be certain extinction. By 1996 there were approximately 40 kestrels in the wild.

The pink pigeon has also been rescued from immediate extinction. The researchers, including Carl Jones and the young ecologist Wendy Strahm, raised pink pigeons in the aviary and then set them out in a small section of forest that was similar to their original habitat. The ecologists were very careful to see that the aviary and the birds' food supply were not contaminated by DDT and other pesticides. 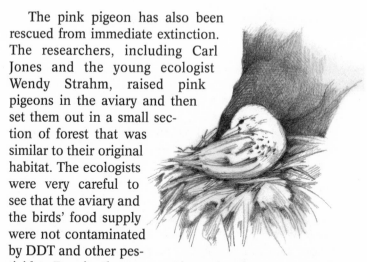 Despite these precautions, the pigeons were not breeding well either in captivity or in the wild. The males did not seem to be interested in sex, and the researchers resorted to artificial insemination.

There are now estimated to be at least 50 pink pigeons alive in Mauritius. The kestrel and pink pigeon populations are probably sufficient to protect the species from immediate extinction. But are they large enough to provide sufficient adaptability to thrive in a rapidly changing environment? Some troubling questions must be answered first. Was the sharp decline in the pink pigeon survival rate due, in part, to inbreeding? If so, artificial insemination may contribute to further deterioration. It will be interesting to see what happens to these species after a number of years have passed.

There are other problems that dwarf the threatened extinction of the kestrel and the pink pigeon. Mauritius is a very beautiful island with dramatic mountains rising from the sea and lush plateaus between the ranges. It is surrounded by transparent waters of a brilliant aqua blue. I was eager to see the underwater life, so I went snorkeling off the beach on the northeast, a place where a number of tourist hotels have been built. It was a very disappointing experience. The coral reefs looked like piles of bleached white bones. The fish population was sparse, lacking in variety and color. When I questioned the local conservationists, I was told that the waters had become polluted with sewage from the hotels and waste from several commercial enterprises, especially the manufacture of textiles and watch dials. These destructive practices have now been recognized and corrective action has been undertaken. One can only hope it is pursued with dedication.

Mauritius has the potential to become the Hawaii of the Indian Ocean—the same spectacular landscape, the same delightful climate—and the peaceful cooperation of four distinct religions and ethnic groups is a model for the rest of the world. But the same pollution problems that plague the Hawaiian chain are also beginning to threaten this paradise. Its physical well-being will depend on how well these problems are controlled.

About a thousand miles to the north of Mauritius lies a chain of islands that have a unique history of formation. The great granite boulders and the sediments that characterize the Seychelles have been dated by radioactive methods and are 700 million years old. These are so ancient that they predate the age of the dinosaurs and the separation of Madagascar and India from Gondwanaland. In fact, they are older by about 300 million years than the first land plants that have been identified in the fossil record.

These little segments of continental crust were once part of India or, perhaps, Madagascar. Although the exact timing is not known, it is believed to have occurred about 66 million years ago. Any life forms that later found their way to these islands enjoyed ideal circumstances for the creation of endemic species.

Julian Huxley once described the Seychelles as a living natural history museum. Even a brief sampling of these islands and the surrounding sea confirms the truth of this statement. Several of the Seychelles have been set apart as nature reserves, and they provide safe havens for some of the rarest and most interesting living things.

Aldabra, for example, the most remote of the islands, is the sanctuary of the flightless white-throated rail, a distant relative of the dodo. It is also one of several islands that are habitats of giant tortoises similar to those on Mauritius. These are implausible creatures, survivors from the time of the dinosaurs, but they did not succumb to the terrible conditions that wiped out their powerful contemporaries. On the tiny Seychelle Islands a greater degree of isolation saved them from the slaughter by human beings that decimated the tortoises on Mauritius and the Galápagos. Slow moving and seemingly lethargic, the tortoises are said to be so large they can crawl carrying the weight of two men. On the islands of Aldabra and Curieuse there are now believed to be as many as 150,000 tortoises.

Aride has the greatest collection of seabirds and a beautiful plant known as *bois citron* (lemon wood), because its flower has an unusual fragrance reminiscent of lemon. Cousin, a beautiful little island, is a protected bird sanctuary supporting abundant avian life. The sanctuary was created to protect the little brown warblers that are the smallest birds on the islands. In 1973 there were only 12, and these were unique to the island of Cousin. A recent census showed that there are now well over 300. They are charming creatures about half the size of one's thumb. They swing on the long fronds of palm trees or morning glory vines as they warble and sway in the breeze.

There are many different kinds of tern living on Cousin. Fairy terns are especially interesting. They are pure white birds that lay their eggs in almost any place, such as on a fallen tree trunk or in the crotch of a tree in plain view. With so little protection from predators, how long could this species have survived in a more competitive environment? On this same island, the lesser noddy tern is so abundant that every tree or bush contains a dozen or so of these birds, and they are remarkably tame. The lesser noddy is another striking example of the viability of exotic species in a habitat where there is very little predation.

An enormous tree that exists nowhere else in the world grows on the island of Praslin in a high, deeply wooded area known as the Valley de Mai. The coco-de-mer (sea coconut) grows to more than 100 feet, and the trees are up to 900 years old. They produce the heaviest seed in the world—double coconuts weighing up to 55 pounds each. Although they float in the ocean and may be carried as far away as the shores of India, they have never sprouted or grown anywhere except on Praslin. When the coconuts were observed floating on the sea and lying on beaches, no one could imagine where they had come from. Perhaps, it was thought, they came from the sea itself, which is why they were called coco-de-mer. On Praslin, the seed does not germinate until it is one year old, and fruit is not produced until the plant is 25 years old. Still, it is mysterious why this one island is the only place in the world where they grow.

A smaller type of palm grows on the edges of the grove at Valley de Mai. The base of its main stalks are armored by hundreds of long, sharp, black spines. It is believed that the spines are intended to protect these smaller palms from the giant tortoises, even though there are no tortoises on Praslin today, but perhaps there were at one time.

Many questions and many mysteries intrigue the traveler to these unsuspected isles set in a far-off sea. They are the perfect place for Alice—a Wonderland where on the tiniest scraps of land trees and turtles grow to be giants, where nuts grow too hard to crack, and birds "grow down" smaller and smaller until they can swing on a morning glory vine. And all of this is caused by selection of the fittest.

"One can't believe impossible things," said Alice.

"I dare say you haven't had enough practice," suggested the White Queen.

The islands of the Indian Ocean are the perfect place to practice suspending our disbelief and, like the White Queen, learn to believe six impossible things before breakfast.

Chapter 8

SRI LANKA

The Resplendent Land

Much have I travell'd in the realms of gold
And many goodly states and kingdoms seen.

—JOHN KEATS

When India separated from Madagascar and traveled at a record rate across the Indian Ocean, it carried with it a fragment of land that later became the island of Sri Lanka. A pearl-shaped pendant hanging by a chain of tiny islets from the southern coast of India, it is an island endowed with beauty and wealth far exceeding that of most lands. Coral beaches ring its shores, coconut groves and rice paddies clothe its lowlands, and cool mountains crown its center, providing the perfect conditions for mile after verdant mile of tea plantations. Pineapples, bananas, breadfruit, jackfruit, and cashews grow abundantly on the island. Fish and prawns abound in the ocean and the many lakes. Food is available just for the plucking, and precious gemstones lie scattered in the soil throughout the land.

Sri Lanka is quite different from the typical islands that we have considered. It is volcanic but very old instead of newly born from the sea floor like Hawaii or Mauritius. It is not really isolated because it is close to India, and a chain of tiny islands makes stepping stones between it and the mainland of India across the Gulf of Mannar. This chain is known as Adam's Bridge, because Muslims believe that when Adam and Eve were expelled from the Garden of Eden they went to Sri Lanka. The bridge provided a thoroughfare for flora and fauna migrating from India. Thus, many of the plants and animals in Sri Lanka are also found in India, and endemic and endangered species are rarer there than in Madagascar, for example. The bridge was a convenient highway for mankind, too, and the favorable environment was so seductive that the human impact on the island is a very important part of its character.

The special qualities of Sri Lanka have been reflected in the names bestowed upon it throughout history. The Chinese called it the Land Without Sorrow, the medieval Arab traders Serendib (meaning "fortunate"). When the

Portuguese took over the island early in the sixteenth century, they changed the name to Ceilao ("celestial"), and when control passed to the British in 1796, they anglicized it to Ceylon. In 1948 the island became independent, and the name was changed to Sri Lanka, which in Sinhalese, the official language, means the "resplendent land."

—

The history of Sri Lanka is documented back to the sixth century A.D., when Buddhist monks began to collect and write the *Mahavamsa*. The text was scratched on palm leaf surfaces with a sharp point. Then the leaves were dipped in carbon black and wiped off, leaving carbon only in the scratches. I have seen pages of this book and was impressed by the fact that the "manuscript" looked remarkably fresh after more than 1000 years. The writing was clear and distinct, and the pages only slightly yellowed. This ancient book was written in a form of Sanskrit from which the present Sinhalese language was derived. The alphabet has 56 letters, all quite similar looking, with circular shapes. Some of the information in *Mahavamsa* is obviously mythical or legendary, but it also contains a substantial amount of verifiable fact.

The Veddas and other primitive tribes were already established on Sri Lanka when the Indo-Aryan Sinhalese from northern India invaded it around 500 B.C. The Sinhalese created city-kingdoms that thrived in the north central plains. This part of the island is very dry except in the summer months, when monsoon winds flow from west to east, bringing in air laden with moisture from the Indian Ocean. During the rest of the year, the winds blow from northeast to southwest, and then the rains come to the other side of the island. In order to equalize the flow of moisture, the Sinhalese built reservoirs and sophisticated irrigation canals, which made this northern plain a very rich agricultural region that supported a highly developed civilization in lavish and extensive cities.

In the third century B.C., the Indian emperor Ashoka, who was a devout Buddhist, sent his son Mahinda to Sri Lanka to convert the Sinhalese monarch. The meeting is said to have taken place on the mountain of Mihintale, which has been honored ever since as a holy place. After converting the king, Mahinda promised to send him several sacred relics, including a right eyetooth of Buddha, a footprint, and a branch of the sacred bo tree under which Buddha was sitting when he attained enlightenment. The branch was planted in the ancient capital city of Anuradhapura. Today, a great tree, which is believed to be the descendent of the original plant, can be seen in the ruins of the great city. Reports of several nineteenth-century explorers who visited the ruins, as well as the *Mahavamsa*, described the evidence of the astonishing wealth and size of Mihintale.

"The sacred bo-tree that stands in the center of a large elevated enclosure," wrote Major Forbes, an Englishman who visited the island and described it in a book published in 1840, "is the principal object of veneration to the numerous pilgrims who annually visit Anuradhapura." Forbes also described the Brazen Palace, with its "sixteen hundred stone pillars placed in forty parallel lines." The *Mahavamsa* says:

> This hall was supported on golden pillars, representing lions and other animals . . . and was ornamented with festoons of pearls all around. Exactly in the middle of this hall, which was adorned with the seven treasures, there was a beautiful and enchanting ivory throne. On one side of this throne there was the emblem of the sun in gold; on the

other the moon in silver; and on the third the stars in pearls. ... The building was covered with brazen tiles; hence it acquired the name of Brazen Palace.

Even allowing for some inaccuracies and exaggeration, the message comes through that this capital city was arrayed in great luxury and extravagance. Lavish shrines, or dagobas, were built to house the sacred relics of Buddhism. They were shaped like enormous beehives and were constructed of earth and brick. One of these, built in Anuradhapura to enshrine the sacred footprint of Buddha, was known as the gold-dust dagoba. It was said to be 270 feet high and 294 feet in diameter—as large as the smaller pyramids of Egypt. These great architectural achievements were begun at a time roughly contemporaneous with the conquests of Alexander the Great and continued throughout the period of the Roman Empire. If we can believe the reports, these palaces and shrines were comparable with the greatest structures built by the Romans.

The remains of several other great cities have been found, including the remarkable fortress-palace Sigiriya, which was built in the sixth century A.D. on a volcanic neck rising 600 feet above the flat plains. This fortress was thought to be impregnable, but the king who built it was finally defeated and the palace was turned into a Buddhist monastery along with its beautiful erotic frescoes. (Our guide remarked that these must have afforded the monks many hours of vicarious pleasure.)

One important shrine of much later construction is still visited by many faithful as well as travelers from around the world. This is the Temple of the Tooth (Buddha's eyetooth) in Kandy, which served as the capital of Ceylon from 1592 to 1815. It was in the Sacred Library of the Temple of the Tooth that I saw the manuscript pages of the *Mahavamsa*.

Hundreds of statues of Buddha decorate this land. The enormous figures stand in solemn rows. The endless repetition, like a chant, proclaims the paths to the good life: right views, right aspirations, right speech, right conduct, right mode of livelihood, right effort, right mindfulness, and right rapture. The overwhelming size of the statues conveys authority and power. A child standing beside one of these may be no higher than the Buddha's big toe.

The civilization that produced these remarkable symbols of faith and wealth was constantly threatened and finally destroyed by invasions of the Tamils, Hindu tribes from southern India. Although these people were always a small minority compared with the Sinhalese, they were tough and courageous. Several times they succeeded in defeating the emperors of the Sinhalese and destroying the works of art in the cities. After centuries of warfare, the buildings and the elaborate irrigation system was demolished or impaired by neglect. The jungle flowed back over the gardened cities. Even today the conflict between the Tamils and the Sinhalese continues. War has broken out again and threatens the stability of the island society.

—

When I visited Sri Lanka in March 1979, a period of peace reigned and the harmonious blend of great natural beauty and the artistic, colorful culture made Sri Lanka seem indeed like a resplendent land.

In many ways Sri Lanka is unusually blessed by nature. Lying just a few degrees north of the equator, the climate is warm and favorable all the year. The lightest shelter is adequate; small, simple native huts are made of woven palm fronds. Even on the beaches these fragile houses provide sufficient protection, because (unlike Mauritius, for example) violent storms and typhoons are rare in these equatorial waters.

The most beautiful portion of the island is the high mountain country where tea is grown. The weather is cooler there than on the coast, and short afternoon showers bring sufficient rain to feed the rushing rivers and many waterfalls. Tea plantations cover mile after mile of this fertile mountain land. The British first planted coffee in the central highlands, but a leaf disease destroyed the coffee crop in the 1860s. Tea was planted as a substitute, and Sri Lanka now produces some of the finest tea in the world. More than 100 varieties are grown here.

I remember especially a view out over range after range of mountains, all covered with the lush shining green of the tea plantations, the dark leaves reflecting the brilliant sunshine. The scene gave the effect of a vast landscaped park. The green of the tea plants was set off by colorful saris worn by corps of native women who moved slowly among the rows, picking the tenderest new leaves of the plants and tossing them over their shoulders into the large woven baskets they carried on their backs. Behind them, like a theatrical backdrop, cascades of waterfalls created shimmering curtains of light.

The city of Kandy is set in a mountain valley rimmed with hills. It is built on a river, and an artificial lake occupies the center of the city, edged on two sides by lacy white stone parapets. Everywhere there is a great profusion of flowers: bougainvillea, jacaranda, vinca, hibiscus, and flame trees. One very spectacular flowering tree, Tabebruia rosea, is a mass of lavender blossoms that come out before the leaves and are shaped like our catalpa blossoms. During our visit the sidewalks of Kandy were carpeted with layers of these blossoms, which popped when I walked over them, like funkia (plantain lily) blossoms in our garden at home.

Every street was a long colorful procession of many people on foot, on elephants, or riding simple wooden carts with openwork sides pulled by oxen. The married women all wore graceful, gaily colored saris. They often carried a sleeping baby on their shoulder. Young girls wore brightly colored short dresses or school uniforms— white dresses with special school ties. The little boys wore blue shorts and white shirts or long white trousers and shirts. The traditional dress for men in Sri Lanka is a white shirt worn loose over white trousers.

The most striking costumes were the saffron robes of the Buddhist monks. The yellow-orange was set off dramatically against their brown skin and the large black umbrellas that many of them carried for shelter from the sun. The robe, draped across the left shoulder and leaving the right one bare, is worn with conscious grace. The Buddhist monks lead very pleasant lives. They are supported by the community, and although they are supposed to practice celibacy, it is a rare monk who follows this precept.

In the lowlands, rice has been raised since very ancient times. The work to produce this crop is all done by hand in the heat of the equatorial sun. As in Bali, planting, cultivating, weeding, picking, and winnowing are all done by women. Working in long lines, dressed in saris with a cloth draped over the head or wrapped turban-fashion for protection against the sun, they add patterns of color to the green of the rice paddies. This green is the most intense seen anywhere in the world—it is a newborn color, a symbol of life itself.

Coconut palms also grow here in vast numbers along the coastal regions and provide many gifts for the people of Sri Lanka. The milk is a cool, refreshing drink. The meat is a delicious food. The nectar from the blossoms is used to make toddy, a sweet drink, which is distilled to produce aurac, the favorite alcoholic drink. Even the husks of the coconuts are put to good use. They are soaked and allowed to partially decompose and soften. Then they are beaten with wooden clubs until the husks fall apart and a tuft of fibers is left. These fibers are woven into a thin twine known as coir. The leaves of the palm are harvested to make baskets and thatch for the roofs of the native huts. Like the tending of the rice fields and the tea plantations, almost all of this work is performed by women. Some of it is very strenuous, even backbreaking work. Sri Lanka may be a second Garden of Eden for Adam but not for Eve.

There are, however, several jobs demanding special skills that are performed by men, for example, the toddy tapping. In certain coconut groves, the tops of the palm trees have been connected by ropes. Three parallel ropes are stretched from tree to tree—two above and one below, providing a tightrope walkway high above the ground. The toddy tappers swing along these perilous highways. At each tree they set out containers to collect nectar from the blossoms, and they lower the full containers to other workers at the foot of the trees. The palms may be 30 to 40 feet above the ground—so tall and flexible that even the gentle winds that typically blow across the island cause them to sway and the ropes to swing. The toddy tappers must be as surefooted as circus stars.

Another traditionally male occupation on the island would be the envy of most males around the world. Many men sit by the sea and fish all day long. The surrounding waters team with fish of many sizes and varieties, and fish is one of the staples of the local diet, more popular than meat. Approximately 220,000 metric tons are caught each year. Along the south coast, where there are especially favorable harbors, a picturesque sight can be seen. Men wearing colorful loin cloths sit or balance themselves on tall wooden perches that are driven into the ocean bed. These perches raise the fishermen about 5 feet above the water level, and the men sit and fish in comparative comfort hour after hour. Groups of them look like flights of long-legged herons, waiting for the fish to strike their lines. Fishing is also conducted in outrigger canoes, which are stable enough to surmount the waves that break on every shore.

There is a third male occupation, which is less comfortable than fishing but perhaps more exciting—the panning for precious stones. Sri Lanka is one of the world's principal sources of precious gems. They are the ultimate gift of the violent volcanism that shaped this land many millions of years ago. Sizable rubies, sapphires, aquamarines, topaz, garnets, and tourmaline can literally be picked up off the ground in Sri Lanka or panned out of gravel beds. Thus, the former name of Serendib is particularly appropriate, suggesting unexpected and happy discoveries.

The old volcanic mountains that form the central massif of Sri Lanka bear evidence of the very powerful volcanic action and tectonic movement that created the heat and pressure needed for the formation of gems. Swirling lines of light and dark layers mark the ancient cliffs. These metamorphic formations were created by forces great enough to bend solid rock as though it were putty, melting the matter and causing it to recrystallize in new shapes and forms. Great outcroppings of igneous rocks and the necks of several ancient volcanoes make dramatic outlines against the horizon. Gleaming biotite and dark red garnet chips stud the exposed rock surfaces.

In many places there are unusually large crystals of mica and quartz, characteristic of pegmatite formations. These showy deposits are created by a process that causes a concentration of certain minerals. Magma forced up from the hot, soft layer of the Earth beneath the lithosphere begins to cool and forms the familiar solid substances that make up volcanic rocks. The minerals that solidify first are those that crystallize at high temperatures, leaving the remaining liquid enriched in those minerals that crystallize at lower temperatures. As this sorting process continues, different types of rock take shape, first the dark basaltic rocks, then the paler granitic rocks. The final liquid residue is a solution enriched in the light elements. When this solidifies in enclosed, protected places, it forms crystals of unusual size and composition. These pegmatite deposits are found in igneous rocks throughout the world. They may range up to thousands of feet in length and hundreds of feet in thickness. A great variety of valuable minerals and gemstones are formed in pegmatites, including mica, tourmaline, quartz, topaz, beryl, and many others.

Occasionally, spectacular crystals of beryl are formed, and they may be tinted in a number of different hues—violet, gold, pink, and many shades of green. The clear blue-green specimens, known as aquamarines, are the most highly prized. Because they are frequently formed in pegmatites, very large single crystals can be found. The limpid aqua coloring of this beautiful gemstone is caused by small amounts of iron incorporated into the crystal structure. Emeralds also are a form of beryl; in this case, chromium is responsible for the deep green color. But emeralds are not found in Sri Lanka, nor are diamonds or gold. (The gold that was so lavishly used in the early Sinhalese cities probably came from India, where there are extensive deposits.)

Some gems are formed in rock that is rich in aluminum. When it is heated by contact with erupting magma, the aluminum fraction separates out in liquid drops and may crystallize as gem-quality corundum (an aluminum oxide). Depending on the coloring agent, spectacular rubies or sapphires may be formed. Many glorious star sapphires with silky six-rayed streaks have been discovered in Sri Lanka and are the special pride of this country. Because these form in a relatively restricted space, they are usually not as large as those that occur in pegmatite deposits, but they are exceptionally pure and beautiful.

According to an ancient Asian myth, the Earth is carried on an enormous sapphire—the celestial stone—and the reflected light from this gem creates the sky. When the Earth is seen today from space, the view is strangely reminiscent of this ancient myth. The Earth looks like an uncut sapphire displayed against the velvet black of space—soft translucent blue with cloudy surfaces etched by the abrasion of billions of years across its rounded face.

The abundant rainfall on the mountains of Sri Lanka has caused weathering of the rocks. This slow process, taking place year after year, century after century, has gradually dissolved the softer materials, leaving the more resistant kernels intact. As the sediments are washed downstream, another sorting action takes place. The heavier particles tend to fall toward the streambed and are separated from the lighter ones that float near the top. Eventually, they are deposited in discrete layers along the river deltas. The lighter layer results in the production of a nutritious red clay soil that is ideal for the culture of tea and other agricultural products. The heavier layer, called "illam," contains assorted rock fragments and pebbles, some of which may be precious gemstones. Swift-flowing rivers carry this load of sediment down into the valleys and spread it out into fans of sediment. In some places the illam may be close to the Earth's surface; in others it may be 20 or even 40 feet deep beneath the clay. Not surprisingly, it is in river valleys that the richest deposits of precious stones have been found.

Ratnapura, the "city of gems," lies in such a valley nestled against densely vegetated hills on the edge of land where rice has been grown for untold generations. Beneath the layer of red clay lies the bed of illam. It looks like quite ordinary gravel with many quartz and granite pebbles; only an expert can perceive that it contains an array of gemstones.

The mining of these treasures is performed in a very primitive manner, virtually unchanged since the mines were first worked thousands of years ago. A hole is dug to the depth of the illam, and the sides of the pit are reinforced with long vertical poles driven against the walls. Then branches and palm leaves are wedged behind them to hold back the soft clay. When water begins to collect in the bottom of the hole, it is bailed out to prevent flooding. A water pump enables the miners to go to depths of 40 feet, but the miners often work knee deep in muddy water. High humidity and tropical heat are intensified in the stagnant air at the bottom of the pit. The men who work there wear only loincloths and a heavy spattering of mud. The mining proceeds at a leisurely pace with an undercurrent of excitement and anticipation, making it seem more like a sideshow than a serious undertaking. Children, relatives, and friends stop by to watch, hoping to witness a moment of great discovery.

The men in the pit fill their round bamboo baskets with gravel, dip them in water, and swirl them expertly, causing the lighter sand, clay, and mud to float to the top and spill over the edge. The heavier pebbles, including any gemstones, collect on the bottom. Periodically the baskets are raised on pulleys to the surface where an experienced sorter examines the stones and sets aside any that look promising. The baskets are then emptied and lowered again into the pit.

Watching this primitive procedure, one finds it difficult to imagine that many precious stones can be found
in this way. Even the best of the pebbles picked out by
the sorter look too dull to be potential jewels. Their surfaces are cloudy and diffused, obscuring the radiant
color that may lie underneath. But each year gems worth
hundreds of thousands of dollars are recovered in these
mines.

The day that I visited Ratnapura one small rough sapphire was found. All work stopped while this stone was
inspected. The miners and the audience crowded
around to inspect the find. When the mud was wiped off,
a cloudy surface was revealed with a hint of translucent
blue. But its potential was hidden beneath an abraded
surface. Only an expert could assess its value.

The precious deposits buried in the crust of the Earth
are retrieved by two different mining methods: deep
mining, in which excavations lay bare the mother lode
(the hard rocks where the stones are embedded), and
placer mining, the removal of minerals from deposits in
sand, gravel, clay , or silt—like the method used in
Ratnapura. Placer mining can be exploited with primitive techniques and, surprisingly, is often the best and
most economical way of mining gemstones. Nature has
already separated the gems from the matrix. It has
washed and tumbled them, sorted them into fractions of
similar size and shape, and deposited them in layers that
are relatively accessible. The placer mining in Sri Lanka
is environmentally benign; it is not threatening the beautiful surfaces that clothe this island with many shades of
green—forests of palm, teak, and sandalwood, rice fields,
and tea plantations.

On many of the islands we visited, special treasures
have been discovered: periwinkle and traveler's palm in
Madagascar, cloves and nutmeg in the Spice Islands,
and sapphires and rubies in Sri Lanka. They are all miracles wrought by the creative forces of nature.

As we look into the heart of the jewel we are brought face to face with a question that has been posed many times but never satisfactorily answered. By what magic does nature draw together the minute bits of precious elements that have been widely dispersed throughout the rocks and waters of the Earth? How does it concentrate them and arrange them in the remarkably pure form in which they are found buried in the mud of Sri Lanka? As the hot, rich solutions containing atoms of many elements percolate slowly through the rocks, temperatures and pressure gradually drop. Then, suddenly, as though touched by the philosopher's stone, molecules of one kind leap from the formless liquid state and, joining together, build that elaborately designed construction in space—the crystal. The presence of just a single crystal or particle of dust can start the transformation of formlessness to form. Atoms of oxygen and corundum or aluminum line up in precisely ordered patterns, like dancers taking their places in a quadrille when the band strikes up. But the speed and the scale of the crystal formation surpass any examples in our human experience. The growing of a typical crystal requires the proper placement of something like 16 trillion atoms an hour.

It is this act of crystallization that has concentrated and laid down rich hordes of treasure in the Earth. In fact, the entire lithosphere is an intricately interwoven fabric of many crystals: rocks and sand, diamonds and sapphires, emeralds and ice. Crystallization is the most important single process that creates the world we know. What a strange irony that from violence and terror emerge order and beauty. The ultimate result is this limpid and quiet perfection that I can hold in the hollow of my hand. It is an expression of the supremely logical structure underlying all things—an orderliness that emerges magically from even the most tempestuously violent forces of nature. As Loren Eiseley said, "It is an apparition from that mysterious shadow world beyond nature, that final world, which contains—if anything contains—the explanation of men and catfish and green leaves."

Chapter 9

CRETE AND SANTORIN

Islands of Destiny

Almost from the moment of its creation, a volcanic island is foredoomed to destruction. It has in itself the seeds of its own dissolution.

—RACHEL CARSON

For many centuries, the island of Crete was in the right place at the right time. It was strategically placed in the Mediterranean—the cradle of the greatest early civilizations of mankind. It was endowed with many other advantages, including a gentle climate and a beautiful and varied landscape from mountain peaks, frequently capped with snow, to warm fertile valleys from which the swallows did not fly in winter.

Girded by the sea, Crete had a natural moat protecting it from attack, and the geography of the island made it especially suitable for a seafaring people. Sheltered bays along the coastline provided favorable harbors. In an epoch when the competing powers of Egypt and Mesopotamia were building only troughlike vessels suitable for river navigation, the shipbuilders of Crete were using keels, creating stable and sturdy ships. Control of the ocean thus became possible for the adventuresome and capable people who made this island their home, and they established a lucrative trade with Egypt and Greece—perhaps also with the Hittites, the Phoenicians, the Babylonians, and even the Far East.

There is, however, one vulnerable aspect of Crete's location that could not have been foreseen by the people who settled there. It lies on a boundary between two lithospheric plates and is the site of unusual tectonic activity. Unlike the Hawaiian Islands, which are situated near the center of the large Pacific plate, Crete lies in a place where two major plates collide—the Eurasian and the African. Over the centuries, the African plate has moved northward, squeezing the Mediterranean Sea, and some oceanic material has been forced beneath the margin of the Eurasian plate. This movement has caused a zone of crustal fracture and volcanism—a fact of utmost importance in the history of Crete.

Although earthquakes did occasionally occur, this did not deter the brave, industrious people who built their remarkable civilization on Crete. After each quake they restored their cities and went on with their lives.

—

The island of Crete was heavily wooded with cypress, oak, fir, and cedar—all popular sources for building materials. It was blessed with fertility at different altitudes, thus providing favorable locations for growing wheat, grapes, and olives, items that were in great demand in the ancient world. On lower slopes of the mountains, where forests had been cleared, olive groves were planted and flourished. Grape vines grew well in the sunny valleys, as did wheat in the flat uplands.

The technique of making bronze—a strong, stable metal—by combining copper with tin, had been discovered and ushered in the Bronze Age. The fleets of sailing vessels that set forth from Crete carried bronze objects as well as lumber, oil, grain, and wine—desirable products to trade in the Mediterranean world.

Works of artistic merit were also offered for barter: jewelry, beautifully wrought of amber and ivory and gold, and pottery, gracefully shaped and decorated with designs of nature that were formalized into elegant patterns.

The precise antecedents of these lively, intelligent, and gifted people are unknown. Very little was known about them until the Englishman Sir Arthur Evans excavated the ruins at Knossos at the beginning of the twentieth century. The treasures uncovered there were more extensive and more significant than anyone had imagined. They revealed a civilization spectacularly unlike any that had gone before.

Some authorities believe that the people who settled in Crete came from Asia Minor as early as 3000 B.C., and they brought with them a Neolithic civilization that was already well developed. Their nearest neighbors, the Greeks, were larger in body, lighter-skinned, and had blonder hair. The Minoans, as they have been called, were little and lithe. They were delicately built with narrow waists and broad shoulders. They had dark, curly hair and darkish skin, bronzed by the sun.

The lifestyle of the upper classes was luxurious and even elegant. They built spacious palaces decorated with colorful frescoes and equipped with amenities like running water, baths, sewers, and toilets with arrangements for flushing. Even the smaller houses were two or three stories high, with wide windows, courts, and often several kitchens. The upper classes had stylish clothes accented with many pieces of jewelry. The women wore a distinctive costume—flounced skirts of several tiers with tight-fitting jackets that came just to the waist and to the elbows but left the breasts completely exposed.

The Minoans invented a form of writing that in its earliest form—known as Linear A—was a highly pictorial writing, similar to Egyptian hieroglyphics. The later form—Linear B—was a variation of archaic Greek, and it was not deciphered until 1952. Both of these styles of writing, however, were cumbersome. They were not good vehicles for creative work but were useful only as a practical way of keeping records and accounts.

The full flowering of the Minoan culture occurred between 1900 and 1400 B.C. In many ways it was unusual for that period in the Earth's history. It appeared to be a peaceful way of life with an enlightened ruler. Their cities were not fortified, and there is no evidence of extensive military activity. We can conclude that the Minoans did not seek to expand their influence by conquest, as did most of the contemporary civilizations. They did not protect themselves with large standing armies but relied on the natural protection of their island position.

The Minoans seem to have been naturally graceful and well-coordinated; these special traits were admired and encouraged. Dancing was a popular form of entertainment for which the Minoans were famous. Acrobatic ability also was carried to a high level of perfection. The art of bull leaping, for example, was an unusual rite, apparently invented and perfected there. Bulls, which were wild on the island, were captured (an act that in itself required skill and daring). Then, young men and young women were sent into a ring to perform dangerous acrobatic feats with these bulls. They would grasp the horns of a bull and somersault over the animal's back or leap onto it and then off, to be caught by another acrobat waiting by the bull's feet. This art may seem to us to have been a barbaric sport, because it could end in a painful and bloody death, but it is not too different from the modern sport of gymnastics, where young girls and boys are trained to perform demanding acts of acrobatic skill. One poorly executed leap could result in a permanent injury. The use of the bull, of course, added an element of suspense and danger. Bull leaping was a test of bravery as well as of grace and skill. And most human societies have invented ways of demonstrating courage, a practice that prepares the young adults to lead the community in a dangerous world.

As it turned out, the most serious danger for the Minoans came not from other competing civilizations but from natural forces deep within the Earth. The sharp, shocking jolt of an earthquake frequently interrupted their pleasant lives, and it was often accompanied by a dull sound rising from the ground, like the muffled roar of an angry bull. Thus, the bull was perceived as a sign of the force that threatened them, the whim of their goddess, the Earth Mother. Homer said, "In bulls does the Earth-shaker delight." To propitiate this goddess, the Minoans sacrificed live bulls. There is no evidence that they practiced human sacrifice, unless the bull leaping is interpreted as such a rite, when they pitted their human

wits and agility against the bull and thus symbolically overcame the force that shook the Earth.

But, as the early Hawaiians discovered with their goddess Pelé, the sacrifices and the rituals did not alter the course of nature. It is estimated that, on average, Crete suffered three serious shocks per century. Archaeologists have identified the fact that about 1700 B.C. the palace of Knossos was destroyed, but it was entirely rebuilt in even greater splendor. There is abundant evidence that the Minoan people were tough of spirit and hardworking. With determination, they overcame the disasters that swept over them.

Eventually something happened to undermine and finally to destroy this spectacular culture. At about the middle of the fifteenth century B.C., its strength sharply declined and there were indications that Greek—or specifically Mycenaean—influence had begun to take over. What could have precipitated this end to the Minoan culture? Was it caused by war or plague or by drought and widespread crop failure? For several decades historians debated and discussed these possibilities, but at the same time geologists and archaeologists were beginning to piece together information that has led to another possible explanation of the decline of the Minoan culture.

Excavations showed that Minoan people had spread to occupy at least one of the neighboring islands. Known as Stronghyli (meaning "round" and describing its circular shape), this island lies 70 miles north of Crete. Cities and palaces endowed with luxuries similar to those at Knossos and Phaistos on Crete had been built on Stronghyli. The remains of these buildings were found buried in volcanic ash and tephra. Cores of deep-sea sediments in the eastern Mediterranean revealed widespread buried ash layers that became thicker toward the location of ancient Stronghyli, their obvious source. These studies yielded convincing evidence of a catastrophic volcanic eruption—an eruption whose violence may never have been equaled in human memory.

In the middle of the fifteenth century B.C., about the year 1456 (a radiocarbon date obtained from a tree that was growing on Stronghyli at the time of the eruption), the people of Stronghyli felt a number of small earthquakes. They were not particularly alarmed at first, because earthquakes had been a common part of their experience, as they were on many Aegean and Indonesian islands. This time, however, the frequency and severity rapidly increased. The more prudent of the inhabitants packed up and moved out, taking their valued possessions with them, but most stayed on until further evidence of impending disaster began to accumulate. Shepherds tending their flocks on the mountainside reported that foul-smelling vapor and steam were being emitted from fissures in the ground and eerie flickering lights were observed at night. At that point, almost everyone wanted to leave, but before the whole populace could be evacuated, the mountaintop burst into flames. A plume like a gigantic mushroom formed and rapidly descended toward the town, carrying chunks of pumice and tephra.* By day the sunlight was dimmed and by night a fiery cloud lit up the sky. General panic and exodus began but there was no room on the ships to take them all. The lumps of pumice were accumulating in the bay and impeded navigation. The last people who sought a place on the boats may have died in the choking darkness.

We can imagine the fear inspired by these sights and sounds on the island of Crete just 70 miles downwind of Stronghyli. The roaring of the bull had become almost continuous. The ground shook and clouds obscured the sunlight, turning day into night for many hours. Ash and solid pieces of ejecta were strewn across the central and eastern portions of the island. Roofs and walls, weakened by earthquakes, collapsed under the load of debris falling from the sky. It was apparent that the Earth Mother was profoundly disturbed. The Cretan people flocked to the sanctuaries and sacrifices were offered. There is even evidence that human sacrifices were performed as a last desperate measure.

The climax of the eruption was so terrifying that it seemed like the end of the world, especially in places so far away that the source and nature of this holocaust could not be identified. The mystery added to the terror.

When the cloud cleared, passengers in ships at sea could see that the whole top of the mountain at Stronghyli had been sliced off as if with a sword. Sailors from ships in the area explored the site and found that the city of Akrotiri had entirely disappeared, covered deep in ash. The truncated cone, which seemed to be

---

*Pumice is volcanic rock that cools too rapidly to crystallize. The vapors dissolved in it, when suddenly released, cause it to foam up into a froth, thus making the hardened rock spongelike and very light. Tephra is a rock consisting of shards of glass that formed when gas frothed through hot lava.

flat, was actually the rim of a large bowl-shaped depression filled with steaming black lava. Frequently, small avalanches of rock and tephra broke loose and slid down the sides of the caldera.

The ruins of Akrotiri were excavated in 1967 by a team under Greece's inspector general of antiquities, Spyridon Marinatos. With several other experts, he made comparative studies between this eruption and that of Krakatoa. By their estimate, the eruption of Stronghyli was four times greater. Marinatos also suggested that this volcanic eruption was the event that brought the great Minoan culture to a sudden end. This explanation was quite generally accepted until later excavations discovered pieces of beautifully executed Late Minoan pottery, which have been precisely dated to several decades after the eruption. If the Minoan civilization was wiped out, who made these works of art?

These findings and evidence of increasing influence from Mycenae suggest that the eruption did not produce an abrupt end to Minoan civilization but only weakened it by damaging the farmlands and the sailing fleet. Then the armies from Mycenae were able to conquer the Minoans.

Using this revised theory, we can reconstruct the immediate effect on Crete and the aftermath of the cataclysm. It is estimated that at least 4 inches of volcanic ash fell on the central and eastern portions of the island, smothering the crops. Heavy and unusual showers, triggered by the explosion, washed ash from the mountainsides down into the valleys and buried even more deeply the crops that had been growing there. Food became scarce and famine spread over the land.

Recovery took place very slowly in the next few years, as the ash was worked into the soil or washed into the sea. But the aftereffects of the eruption continued to take place, and these were devastating to the Minoan people, who were demoralized and weakened by privation and hunger.

Back at the site of the eruption, the great caldera at Stronghyli slowly sank lower and lower, until it was not much higher than sea level. Then, one day a major piece of the rim collapsed. The sea rushed in, filling the bowl-shaped crater, making a bay where the highest mountain peak had once stood. This sudden movement created a tsunami, which moved outward with lightning speed through the Aegean.

Imagine such a wave descending on the shores of Crete—a civilization dependent on its commerce and its control of the sea. Its harbors were filled with shipping and sheltered by arms of the land that served only to funnel the power of the wave. We can recapture the scene with details gleaned from tsunami events in more recent times.

The first visible sign that a tsunami is approaching is a withdrawal of the water like a rapidly falling tide reaching levels far below the lowest neap tide. The edge of the sea may retreat to several miles offshore, baring

the muddy ocean bottom and leaving ships stranded on the naked shore. Anything loose, including people, can be drawn out to sea. When the waters return—sometimes as long as half an hour later—they come with such force that they can pick up boats weighing tons, lift them over piers and houses, and disgorge them 100 feet or more inland. Then, when the danger might appear to have passed, people may flock to the shores to assess the damage. And the next wave of the tsunami descends (a typical one comes in several waves, one of which may be higher than the original).

As pieces of the old caldera at Stronghyli continued to crumble, more of these great waves were generated. In addition to the physical destruction, they caused psychological damage, because they descended without warning and no one understood the cause. Fear grew in the hearts of the Minoan people. It was apparent that the gods had singled them out for some special punishment. Even sacrifices of their most honored possessions did not affect the course of nature.

Throughout these years more earthquakes also shook the land, especially in the eastern part of the island. Finally, one night the gods struck out with decisive fury. A major earthquake centered deep under the city of Zákros in eastern Crete shook the entire island. The mud-brick, multistory buildings rocked and toppled like houses built of toy blocks. Oil lamps that just a few minutes earlier had cast their mellow glow on rich banquet scenes and luxurious interiors were overturned. The holocaust of fire was added to the din of collapsing masonry and the screams of victims caught in its path. The fire was so intense that large pieces of adobe brick were partially melted. Archaeologists have discovered that many of the other major towns in Crete were severely damaged, and walls were charred by fire.

It is easy to understand how survivors of this series of disasters did not have the heart to rebuild as their fathers and grandfathers had done so many times before. Better to leave this cursed land and go on to other places that were in higher favor with the gods. Those who could afford to do so picked up what was left of their possessions and moved on. The Minoan nation, depleted of its most successful citizens, never recovered its strength and power. Many of the refugees settled in Mycenae on the mainland of Greece, bringing with them a high cultural tradition that contributed to the emergence of the Greek culture and through it the entire Western world.

The shape of the nearby islands was also altered beyond recognition. The single round island of Stronghyli had disappeared and in its place stood five small islands. The group is now known as Santorin and the largest island as Thera, although this distinction is not always observed.

It is reasonable to suppose that the events that created the island group of Santorin and changed the course of Minoan and Greek history might also have had a significant impact on other civilizations around the Mediterranean. Egypt, which lies about 500 miles directly downwind of the site of the volcano, was a place that could have felt the influence of these events. In fact, there is some evidence that here, too, the eruption had a significant impact on human history.

Egyptian writings of approximately 1500 B.C., during the Eighteenth Dynasty, recorded the fact that imports of Cretan cedar and olive oil needed for preparing mummies had ceased about that time. Those same writings tell of floods and days of darkness when "the sun appeared in the sky like the moon."

It has been suggested that these awe-inspiring events are similar to the biblical story of the events leading up to the Exodus—the time when the tribes of Israel departed from the land of Egypt. According to the Bible, alarming and unexplained punishments were inflicted on Egypt: There was sudden darkness lasting for several days, the waters of the river turned to blood, the fish died, the frogs came up and covered the land, all the cattle died, and the dust of the earth caused boils and blains to break out on man and beast. "The Lord sent thunder and hail; and the fire ran along upon the ground. . . . And the hail smote throughout all the land of Egypt . . . and the hail smote every herb of the field and brake every tree of the field."

Many of these phenomena can be interpreted as consequences of volcanic eruptions or earthquakes. When a cloud of ash blacks out the sun, the weather is usually affected, bringing unusual amounts of rain, thunder, and lightning. Hail also is commonly associated with an ash fall, even quite far away from the eruption. The ash particles act as nuclei around which ice can form. Swarms of insects have been observed in the wake of heavy rains that cause an unusual growth of desert vegetation. In some cases volcanic ash is contaminated by traces of fluorine, which is corrosive and poisonous, causing the death of livestock that graze on grass dusted with this ash. Following major volcanic eruptions, human hygiene has broken down, because drinking water has been polluted and diseases are carried by insects. Ash contaminated with fluorine may cause boils and irritation on exposed skin.

Several of the phenomena reported in the Bible are suggestive of reports received from other times and other parts of the world. The Chinese have observed that during earthquakes frogs jump out of ponds and rivers, the water of the wells turns the rice red, the waters froth and are evil-smelling, and animals sense that something is awry (the panda moans and holds its head in its paws).

The most striking resemblance to earthquake phenomena is the parting of the Red Sea. As previously mentioned, the approach of a tsunami is usually characterized by a dramatic withdrawal of the water in a bay adjacent to the sea. And this is followed—perhaps half an hour later—by the return of the sea to levels much higher than the normal level of the bay. The Red Sea itself lies too far from the shore of the Mediterranean to have suffered this change in sea level. However, the suggestion has been made that the body of water described in the Bible was not the Red Sea but a shallow lagoon lying between the Nile delta and the sea. A literal translation of the name *Jam Suf* from the Hebrew original is Reed Sea, not Red Sea, and because reeds do not grow in saltwater, the theory has been suggested that the tribes of Israel traversed one of the small bodies of water at the edge of the Nile delta. These would have been exposed to the full forces of a tsunami caused by the eruption at Stronghyli. The waters retreating before the approaching wave would have bared the bottom of this lagoon, thus permitting a dry passage for the Israelites. Later, the returning waters would have drowned the pursuing forces of the pharaoh.

There is still considerable controversy about the validity of this explanation—scholars have questioned the precise date of the Exodus, the identification of *Jam Suf*, the size of this body of water, and the length of time required to allow passage of the Israelites. However, it does present a very interesting possibility suggested by geologic facts. We recognize that the eruption at Santorin was a happening of such major significance that it could have been responsible for many of the strange, alarming manifestations that persuaded the pharaoh to allow the children of Israel to depart from the land of Egypt.

In any case, the volcanic eruption at Stronghyli changed the whole course of human history. It redirected the flow of Minoan civilization into the mainstream of Western culture. Perhaps, too, it freed the enslaved tribes of Israel, resulting in a new direction of thought and belief, creating a distinctive lifestyle that was destined to have a significant impact on mankind.

Today the little island of Santorin dreams in the sunshine. The whitewashed houses of the town are set like a crown of pearls high above the azure blue of the Aegean Sea. Below the town, precipitous cliffs descend to the bay where the summit of Stronghyli once rose. The steep ascent and descent can be made on donkey-back. The slow pace and hypnotic rhythm of the donkey's footsteps impart a timeless quality to this trip across the great caldera rim with ancient history written on its fractured face. Layer upon layer of ash, lava, and pumice are quietly folded now, pressed together like flowers in the pages of a book. Only the striking spectrum of colors—deep red, rose, pale pink, chocolate, light buff—suggests the drama recorded here. In this bright dust is the story of days and nights of terror: cities deserted, armies drowned, navies crushed, and the whole course of human culture redirected down new paths.

The Earth here seems to be at rest but it will move again. It will tremble beneath the desert sands of California and the cherry orchards of Japan. These events are inevitable, considering the vitality, the strength, and the power embodied in our planet. The energy involved is past imagining. And man—in his tiny adobe dwelling perched on the edge of the precipice—will be shattered again and again by these tremors until he learns to understand the Earth, to sense its stirrings, to interpret the past, and to foretell the future of this deceptively quiet dust that he can hold cupped in the hollow of his hand. As the poet John Wheelock tells us,

> 'Tis natural yet hardly do I understand—
> Here in the hollow of my hand
> A bit of God Himself I keep,
> Between two vigils fallen asleep.

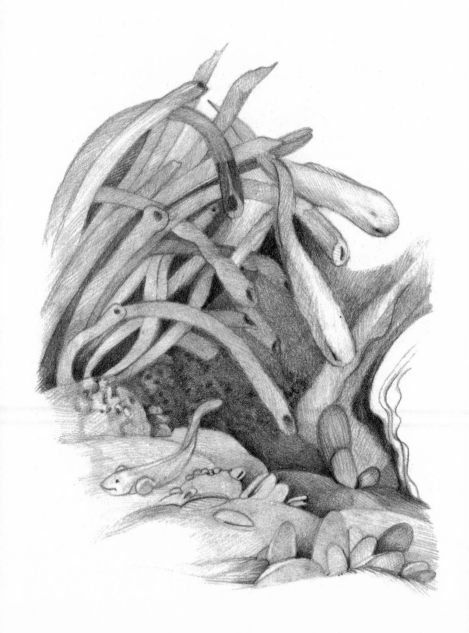

Chapter 10

ICELAND

Crucible of Creation

Time writes no wrinkle on thine azure brow:
Such as creation's dawn beheld . . .

—LORD BYRON

Volcanism is both creator and destroyer. Many of the islands we have visited—the Hawaiian Archipelago, the Galápagos, Mauritius, Easter Island, and the tiny Spice Islands—were born from volcanic action. Actually, all the land masses on Earth, continents as well as islands, owe their existence to the stirrings of this great power in the soft, hot interior of the planet. In the experience of mankind, however, it has seemed to be the very incarnation of an evil force, a voracious beast lying in wait to suddenly spring up and devour whole towns, spreading ribbons of fire on the fragile abodes of men, on green pastures and flowering hills. From the beginning of human history, people have sought to appease this frightening force, such as the sacrifices to Pelé and the statues at Easter Island that faced inward to guard the land. Volcanoes have forced the evacuation of thousands of people and the destruction of civilizations.

And yet we also know that it has brought great gifts, bestowing fertility to our fields and laying down treasures of gold, silver, and gemstones in our mountains. Even the cool blue waters of the ocean and the soft breezes that freshen a summer night were born in the reeking conflagration of volcanic eruptions. In fact, all the atmospheres of the inner planets are believed to have had a volcanic origin.

The Greek philosopher Empedocles, who was born five centuries before Christ, lived in Sicily less than 100 miles from Mount Etna, the largest volcano in Europe. He was perhaps the first person to sense in this terrifying natural phenomenon a creative rather than a destructive force. All things, he said, can be reduced to four primary substances: earth, air, fire, and water. It is the coming together and parting of these elements that make up the great diversity of the physical world.

An illustration of how this happens is vividly demonstrated by the history of Iceland. Only a few people have seen the earth opening up right at their feet and watched the violent process taking place before their eyes. The volcanologists Maurice Krafft and Katia Krafft described such an experience in 1973 on the island of Heimaey in the Vestmann Archipelago south of Iceland:

Between midnight and 1:35 A.M. on January 23, the earth shakes three times on the island of Heimaey. Being so used to this kind of thing, the inhabitants scarcely notice it and go on sleeping peacefully. But two men, two fishermen, are walking in the streets in the eastern part of the town of Vestmannaeyjar. At 1:55 A.M., they suddenly see a line of fire behind the houses; it spreads, noiselessly and rapidly, from south to north. They have been drinking a little, of course. Convinced they are having hallucinations, they go closer to the phenomenon, and then they understand that the line of fire is a gaping fissure that spits a curtain of molten lava more than 100 meters high. A new volcano is being formed! The two men forget to give the alarm; they run home, wake their wives and children, then rush to the harbor. A few seconds later the telephone rings at the police station of the town; the incredible news spreads. The municipal authorities are immediately alerted; the police, blowing their sirens, rush through the streets of the sleeping town; the whole population awakens, jumps out of bed, and looks with astonishment at the growing, reddening volcano. . . . The civil-defense department, the airport of Keflavik, the hospitals, and the police are mobilized without delay. . . . Six hours after the beginning of the eruption, the last inhabitants are evacuated; approximately 5,000 people have been transported to the mainland.

By the end of the next six months the activity had subsided. Three hundred houses disappeared under the lava, and more than 100 were buried under ashes or burned. The island sank slowly because a great void existed underground, which had contained all the magma that emerged from the erupting fissure. Ten days later, the new volcano was given an official name—Eldfell, for "the mountain of fire."

At first the atmosphere in Heimaey contained an unusual mixture of gases; it was heavily laden with carbon dioxide, carbon monoxide, hydrogen sulfide, and methane. Water was also ejected in the volcanic plumes that streamed upward from the eruption and joined the clouds that move in restless patterns around the globe. Most of this water had previously existed in the form of cloud droplets and rain. It had percolated down through the ground, where it had collected in pockets and reservoirs. During the eruption it was ejected again into the air. But some of the water had been removed from chemical combination with the rocks and soil in the heat of the volcanic eruption. This "juvenile" water, which had never been part of the Earth's atmosphere before, may be the original source of all the water present on our planet. When the Earth was young, volcanic activity was very intense and a larger proportion of the water emitted from volcanic eruptions was juvenile water. An untold number of violent events must have split the planet's crust and poured forth streams of gases and water to join the primordial atmosphere. Gradually, an atmosphere and surface water accumulated and evolved.

Geologists believe that a similar process occurred on other planets. There is evidence of extensive volcanism and crustal deformation on Mars, for example. Mars has a volcanic cone that dwarfs any volcano seen on Earth; it is some 350 miles in diameter and 4 times as high as Mount Everest.

In 1986 I visited the town of Vestmannaeyjar on the island of Heimaey. The houses spilled down the valley just to the brink of the sea. The steep jagged outlines of mountains on either side were cloaked in soft black ash, which was still very warm to the touch. In fact, the local people take advantage of this energy by burying pipes carrying water deep under the ash and circulating the warm water to heat their houses.

After leaving the island of Heimaey, our ship passed close to the island of Surtsey, which was created by an eruption in 1963. In November of that year, ships coming into or leaving the harbor at Heimaey saw a huge jet of water rising from the surface of the ocean near the southern coast of the island. Two years later the signs of the eruption had ceased, and an island about 1 square mile in area had materialized. It was christened Surtsey for the god of fire in Icelandic mythology.

When I saw the island, it looked like a gigantic black footprint on the surface of the sea. It was an ominous sight, like the cinders of a conflagration that had burned a piece of the Earth, leaving just a jumbled layer of lava rocks. I was not able to go ashore on Surtsey; access to the island is limited, because it is being carefully monitored for geological and biological research.

The passage between Surtsey and Heimaey was guarded by a series of dark rocks, like black warning fingers rising from the sea,

and interspersed by many icebergs, some of them very large. The captain of our ship had to weave his way carefully among these hazards in heavy seas stirred by a northeasterly gale.

The eruption that created Surtsey, however, was mild compared to the tectonic blast that occurred 20 years later on Iceland, cracking open the huge ice field Vatnajökull in the Öraefa Glacier, the largest in Europe. After a series of telltale earthquakes in early fall 1996, a powerful blast carved an ice canyon 500 feet deep and more than 2 miles long. It split the rocky crust underlying the glacier, and intensely heated magma welled up through the fissure, melting the ice field and producing torrents of water that poured through the canyon until it collected in an icecapped crater lake, known as Grimsvötn Caldera. Billions of gallons of meltwater filled the lake to bursting. Then, just after dawn on November 5, the 800-foot-thick icecap was blown off the lake, and with alarming speed a huge volume of water poured from the lake, cascaded beneath the glacier, creating a river that rivaled the flow of the Earth's second largest river, the Congo. At the edge of the glacier, almost 30 miles away, the floodwater emerged with such enormous force that it broke loose blocks of ice estimated to weigh 1000 tons. These slabs of ice, mixed with dust and sediment, were carried miles downstream and scattered helter-skelter over an immense area. The melting of these slabs produced a field of odd-shaped rocks similar to scenes that have been found in our exploration of Mars.

It is no coincidence that Iceland has provided the model for the volcanic processes that have shaped the

Earth and perhaps the other inner planets of the solar system. This little island lies in one of the positions on Earth where maximum volcanic action is still occurring. In a situation similar to the Galápagos, it sits athwart a giant fracture in the Earth's oceanic crust, the Mid-Atlantic Rift. This spreading center created the Atlantic Ocean during the past 200 million years. The rift is a long jagged split in the ocean floor through which hot magma forces its way up from the deep inside layers of the Earth and pushes apart two of the major plates of the planet's surface—the American and Eurasian plates. The space between them is growing at the rate of 1 to 2 centimeters a year. Iceland is situated directly on this underwater ridge, which stretches all the way from the north to the south polar regions. In fact, the island is actually part of the ridge. One can see on Iceland the largest exposed area on Earth, where the crust of the Earth is being torn open. This prominent structure is conspicuous across the center of the island from north to south and 30 active volcanoes occur along the zone.

Iceland is remarkable for the number of its volcanoes and for the frequency of its earthquakes. On the island as a whole (an area about the size of the state of Maine), there are more than 100 volcanoes, 120 glaciers, and numerous small lakes and swift-flowing rivers. Mount Hekla has erupted many times since records have been kept (most notably in 1766, 1947, and 1980). Nearby Mount Laki has 100 separate craters. In 1973 Laki erupted, pouring out molten lava and volcanic ash and producing torrential floods that led to the deaths of more than 9000 people. Large tracts of arable land were ruined, and about 80 percent of the livestock on the island was lost.

Thermal springs are also common in Iceland. Particularly numerous in the volcanic areas, the springs occur as geysers and as boiling mud lakes. The geyser named Geysit, the most spectacular one, erupts at irregular intervals, ejecting a column of boiling water up to almost 200 feet in height.

The history of the island demonstrates the strength and persistence of the tectonic forces. Under normal conditions, geothermal heat rising beneath Vatnajökull melts enough ice to trigger a flood every few years. Since records have been kept, beginning in the ninth century, 60 catastrophic floods like the one in November 1996 have occurred. The dramatic landscape of this small land mass is being continuously carved by an irresistible force stirring like a restless giant in the mantle of the Earth.

The history of Iceland also provides clues to an understanding of the relationship between the Earth and the sun. Evidence of long-term climate changes has been suggested by the tales embodied in the old Norse sagas and by the remains of Viking settlements in Iceland as well as Greenland, its nearest neighbor, which lies just to the west.

Iceland was settled in the mid-ninth century by adventurous Norsemen sailing from Scandinavia in open longboats. The passage at that time must have been mild enough to make this journey reasonably safe and to encourage other settlers to follow. According to historical records, all of northern Europe enjoyed an unusually benign climate between the years A.D. 1200 and 400 The climate in Iceland was favorable enough for colonization and a stable community was established.

Among the colonists living there was Erik the Red, named for the color of his hair and beard. In 981 he was banished from Iceland for three years as punishment for murdering two men in a feud. Erik had heard a persistent rumor that there was more land beyond the sunset. He was inspired by this tale, and when he left Iceland he sailed west and discovered a large island about 180 miles away, which he named Greenland and is now known to be the largest in the world. There is no mention in the sagas that he encountered ice drifts, which are now common in these northern waters. During his three years of banishment, he explored this land and then returned to Iceland with stories of lush landscapes and fertile pastures on the island.

Greenland was probably never very green, but Erik was a skillful promoter; he believed that more people would follow him to the new land if it had a beautiful name. And at that time Greenland did have a climate benign enough to grow vegetables and grasses to make hay for livestock.

Attracted by these reports, about 300 men, women, and children, together with cows, horses, sheep, and household goods, set sail in open Viking longboats for new homes in Greenland, where they established several colonies. They built houses, cultivated farms, bred cattle, wove cloth, hunted, fished, and traded with Europeans. There seems to have been no difficulty in maintaining regular communication with the continent during this warm period. In fact, more and more Viking families made the long journey from Scandinavia with their possessions to join the colonies, which prospered for several centuries.

In the thirteenth and fourteenth centuries, however, the communities decreased steadily in size. Travel to and from Europe became more difficult. Ships headed for Greenland had to go farther south to avoid the drift ice and then swing back north to reach the southwestern coast settlements. Eventually, all communication with the homeland ceased.

Excavations of the ancient farms and grave sites show that many farms had been abandoned; fewer cattle, sheep, and horses were raised. The human skeletons found in the later graves were small in stature; by about

1400, the average Greenlander was probably less than 5 feet tall. Many seemed to have been severely crippled and twisted. The cause of this condition has not been definitely determined—disease, poor nutrition, genetic inbreeding, or perhaps decalcification of the bones after death.

Most striking of all is the fact that the graves became shallower as time went on. They were 6 feet deep in the twelfth century but gradually decreased in depth, until by the end of the fifteenth century they were barely deep enough to cover the bodies. These facts can be most easily explained as the result of increasingly colder weather, making it more and more difficult to dig into the permanently frozen ground.

I visited the site of one of the last Viking settlements. It was in a protected situation at the end of one of the most southerly fjords on the west coast of Greenland, and it was built on the south-facing hillside, taking maximum advantage of solar heat. The remains of many stone buildings were scattered on this hillside: a church, meetinghouse, cow barn, and foundations of many smaller buildings, probably individual homes. The last written history of Viking occupation of Greenland was the record of a marriage that was celebrated in the church in 1495. The couple then moved to Iceland, taking the record of their marriage with them. In fact, many colonists left the island around this time. The communities rapidly decreased in size, until in the early sixteenth century none was left in Greenland. Today, the site of Erik the Red's settlement is once again green; sheep graze there and the glaciers have melted back to open the fjords. The site of this little community seemed very favorable to me on this warm, sunny, August afternoon, with the early evening light bathing the mountainside in a golden glow.

As our ship sailed south toward the tip of Greenland, we entered the Prins Christian Sound, and there I received a very different impression of Greenland. The sound is a narrow channel with tall, dark mountains rising straight up on either side, almost blocking out the sky. They were dusted with fresh snow and festooned with tumbling glaciers and waterfalls—it was an awesome sight. The ancient rocks that form this land are among the oldest in the world. They are products of the volcanic activity that created the crust of the Earth 3.8 billion years ago.

Erik the Red is believed to have found this passage and used it to avoid passing farther south around the tip of Greenland, in order to reach the more sheltered western coast, where the settlements were built. I thought of him sailing down this dark, dreadful channel not knowing what hazards lay ahead, with glaciers and waterfalls and ancient rocks lying just beneath the surface of the sea, and I realized what a powerful desire for adventure and physical courage drove this man.

These same traits were passed on to his second son, Leif Eriksson, who continued his father's exploration to the west and founded the first little settlement on the continent of North America. He was the true discoverer of the New World.

Leif had heard reports of an Icelandic trader named Bjarni Herjulfsson, who in 981 had been caught in a storm on a passage between Iceland and Greenland. When the storm abated, he made landfall on the coast of a tremendous landmass previously unknown to white men. Although he made no attempt to explore or colonize this land, he carried the tale back to Greenland. Leif was excited by the story, and in about the year 1000, he bought Bjarni's ship and attempted to retrace the voyage, based on Bjarni's description. He sailed along the coasts of what are now known as Labrador and Newfoundland, finally settling down on a little harbor in northern Newfoundland. Like his father, he was able to attract colonists to follow him. Like his father, too, he gave the new location an alluring name—Vinland—because he had found grapes growing there, which seemed to promise a favorable agricultural setting, including a sunny and pleasant climate. But the new settlement did not last long. In 1963 the site named L'Anse aux Meadows was excavated, and anthropologists found remains of the old Viking village. Today, the climate in northern Newfoundland is too severe for the culture of grapes.

—

The story of the Viking adventurers does confirm the fact that the Earth's climate has passed through long periods of change, some lasting for many centuries. The most obvious reason for a significant climatic change is an alteration in the amount of energy received from the sun. At first glance, this possibility seems to be extremely unlikely. Why should the energy radiated by the sun vary at all? Within the spans of our lifetimes any small fluctuations in climate seem to average out from year to year. Actually, it is known that significant long-term climatic changes have occurred, and scientists believe that these variations were caused by differences in the amount of radiation received from the sun. The intensity of solar radiation falling on the Earth's surface varies in a number of ways. The sun passes through cyclical periods of slightly increased or diminished activity approximately every 11 years. This is a relatively minor variation, but there are more significant changes in the effectiveness of the solar radiation caused by cyclical changes in the shape of the planet's orbit, the tilt and precession of its axis as it moves around the sun. These changes on Earth are perhaps sufficient to have precipitated the series of ice ages that are known to have occurred on the planet.

For some time scientists have believed that all life depends exclusively on sunlight for its energy, but it has recently been discovered that life can thrive in situations where the sun's rays do not penetrate. Volcanic action along rifts on the ocean bottom cause springs of extremely hot effluent to boil up from the Earth's mantle. Geysers of water as hot as 700 degrees Fahrenheit shoot up with great force 20 or more feet high. These fountains erupt from the sea floor just as Geysit and other geysers erupt on land.

A weird assortment of aquatic creatures inhabits these underwater vents, clustering thickly in the vicinity of the hot springs: previously unknown species of fish, clams, mussels, lobsters, sponges, sea cucumbers, and worms. The water in the jets is thick with bacteria of a special kind that extract the energy stored in hydrogen sulfide. This smelly compound holds the secret of the abundant life in these regions, which are entirely cut off from the energy of sunlight. Under extreme heat and pressure, the bacteria use the hydrogen sulfide combined with carbon dioxide and oxygen dissolved in the seawater to synthesize the organic compounds that are food for higher life forms.

This unusual life-support system (the only one known that is not dependent on the energy of the sun) is so successful that the living community is profuse, and the species grow at extraordinarily rapid rates compared with other aquatic organisms. Giant clams cluster in a dense blanket on the ocean floor; many of the rocks are covered with enormous mussels. In one area reddish fish of an unknown species spend most of their time with their heads thrust into the sulfur vents, feeding and moving their tails in unison like a flickering red flame. In another location, tube worms 6 feet tall grow tightly clustered together and open their spectacular red, flower-like plumes to comb the nutrient-rich waters. The life here is so profuse that the deep-diving research team in the scientific submersible *Alvin* named one of these hot springs the "Garden of Eden."

From discoveries like this we have learned that life can exist in conditions that would have seemed very unlikely a few years ago. And we recognize that volcanism may provide the conditions in which life can thrive in the dark depths of the sea.

The presence of water was a crucial factor in the early stages of life's development on Earth, because it protected and shielded the vulnerable living organisms from sunlight. Although we have traditionally considered sunlight to be the energy source on which life depends, there is a very dangerous component in the sun's rays that can be lethal to living things. This is one of the reasons why scientists believe that life must have begun in the sea. It was not until after photosynthesis was invented that life could leave this soft, sheltering medium. Through this remarkable process certain forms of living things could utilize the sunlight as a source of energy, removing carbon dioxide from the environment and releasing oxygen as a by-product.

With the accumulation of oxygen, an ozone layer began to form, and ozone has the special characteristic of absorbing the very energetic ultraviolet component of sunlight, thereby shielding fragile organic molecules. When the concentration of ozone reached a safe level, life was able to leave the water medium to invade and populate the land. There it was wrapped by the atmosphere, which had become a sheltering, nurturing medium that tempers the climate and provides many of the chemicals and water that are essential for the sustenance and growth of living things. Thus, the story of life is intimately associated with volcanism, the power that molds new land forms, removes water from the very rocks themselves, and spins the gossamer fabric of clouds and mist and rain.

Indeed, according to one theory, life itself may have been conceived in a volcanic eruption. In and around the dark clouds that accompany eruptions, intense electrical storms rage. The air is electrified for miles around. As the sailors in the Sunda Strait observed during the eruption of Krakatoa, flickering pink flames, which they could not put out with their hands, danced along the masts and spars of the ships, a phenomenon known as St. Elmo's Fire. Perhaps, so the theory goes, the electricity and heat associated with volcanic eruptions provided the energy that synthesized the complex protein molecules that served as the precursors of the simplest living things. Perhaps Empedocles was not far from the truth in suggesting that all reality involves the coming together and parting of earth, air, fire, and water. Thus, the island of Iceland provides important clues to answering that most tantalizing enigma—the mystery of creation.

Chapter 11

THE BAHAMA ISLANDS

Daughters of the Sea

The face of places, and their forms decay;
And that is solid earth, that once was sea;
Seas, in their turn, retreating from the shore,
Make solid land, what ocean was before.

—OVID

The Bahama Islands were born from a shallow, tropical sea that the Spaniards called Baja Mar, meaning "shallow sea." They are made up of sand and the skeletons of ancient coral reefs. Scientists believe that these reefs began to form as long ago as 350 million years. Coral and algae communities began to grow on the windward side of a wide continental shelf made by the erosion of sediments from North America. Westerly ocean currents impinging on the edge of the submerged shelf were deflected upward, bringing fertile deep-sea waters to the surface and creating optimum conditions for these reef-building organisms. The growth was so rapid that reefs rose from the seafloor and soon reached sea level. The windward side of the reef was steep-faced, constantly battered by waves. On the protected leeward side, sands consisting largely of coral fragments were scattered on the existing continental shelf and covered broad areas of shallow water—now known as the Bahama Banks.

It is a fascinating experience to fly over this area and, looking down, see the ocean bottom as clearly as if it were dry land. The water is so astonishingly clear, imparting just a faint blue tinge to the underwater scene where drifts of sand make gently curving hills and ripples march in serried ranks across the valleys.

A river of warm water flows between the Banks and the coast of North America. It is part of a clockwise rotating system of currents in the North Atlantic. As it flows northwest from the West Indies, it is deflected sharply north by the peninsula of Florida and confined by the submerged Bahama Banks. It narrows and becomes swifter, moving at about 4 miles an hour, and it carves out a channel on the ocean bottom, which is 2600 feet deep. Thus, a relatively deep strait separates the Bahama Islands from the coast of Florida. Farther north, the Gulf Stream slows down and follows the continental slope beyond the edge of the shelf. At times of

low sea level—like the last ice age, 12,000 years ago—
the Bahama Banks were exposed dry land, but the
islands were still isolated from North America by the
Florida Strait. There may have been better land connec-
tions with South America, and there seems to be some
evidence that living things traveled that way to settle the
islands, just as Bali (another continental island) received
flora and fauna from Asia.

Birds that inhabit the Bahamas are more closely
related to Central and South American species than
North American species. There are parrots, bullfinches,
brilliant orange tanagers with black stripes, cuckoos,
and banana quits in the islands but no pelicans or cardi-
nals or blue jays.

Our knowledge of the early human inhabitants of
these islands is limited because they had no written language. The best sources of information about them are
the reports by Christopher Columbus to King Ferdinand
and Queen Isabella of Spain, who sponsored and
financed his journey of exploration. Columbus made his
first landfall on one of the Bahama Islands on October
12, 1492. The precise identification of the island is still
under dispute, but most authorities believe it was San
Salvador, which lies close to the center of the archipelago. He found a colorful land of virgin beauty and
friendly natives who called themselves Lucayans. They
were handsome with brown skin and well-proportioned
bodies, a gentle and peace-loving people.

At the time of Columbus's discovery, the Lucayans
occupied the Bahamas, and closely related tribes—the
Arawaks and the Tainos—inhabited Cuba, Jamaica, and
Hispaniola. Like the Lucayans, they were nonviolent and
pleasant people. But the islands to the east and south
were occupied by a cruel, warlike tribe—the Caribs, who
were cannibals. They frequently raided the Bahamas,
arriving secretly by night, carrying off men, women, and
children in canoes. The males were mutilated, tortured,

and eaten. The women were kept for breeding purposes. At any time without warning this terrible danger could descend on the Lucayans from the sea.

In 1492, however, a new danger came out of the sea in tall ships with white wings and men with white skin. The Lucayans were filled with awe and wonder. They welcomed this new visitation with gifts of food and celebration, little suspecting that it would lead to the worst disaster that they would ever suffer. Within 25 years their entire tribe, which numbered 20,000 to 30,000, was wiped from the face of the Earth.

Although Columbus admired the kind, happy dispositions of the Lucayans, he noted that they were very timid and malleable. "They should make good servants," he wrote in his journal. A few days later his mind turned from servitude to slavery. He wrote to the king and queen of Spain from San Salvador: "When your Highnesses so command, they can all be carried off to Castile or held captive in the island itself, since with fifty men they would be all kept in subjection and forced to do what ever may be wished."

Columbus noticed that some of the natives wore gold ornaments, and he was determined to find the source of this wealth. He explored many of the Bahama Islands but found no gold. Pressing on farther south, he visited Cuba and then the beautiful mountainous island he later christened Hispaniola. Unlike the Bahamas, this island was volcanic and did contain veins of gold. The Taino people there were closely related to the Lucayans and similar in disposition.

Within two years the entire island of Hispaniola was turned into a brutal labor camp, with the Tainos forced to work long hours panning the riverbeds and digging in the mountains to locate deposits of gold. They were fed so little that they lived in a state of perpetual hunger and weakness. There were not enough hours in the day to meet the quotas of gold, and for this failure brutal penalties were inflicted on their emaciated bodies. Many did not survive this treatment. It has been estimated that when the Spaniards first came to this island there were 300,000 Tainos living there. Sixteen years later only 60,000 remained. And by 1550 there were probably fewer than 500.

When the Spaniards were faced with a rapidly diminishing labor supply, they turned to the Bahama Islands to restore it. Slave ships visited the Bahamas; with cutlasses and savage dogs the Spaniards drove whole communities of Lucayans aboard these ships. The captives were then crammed into dark, suffocating holds below deck. With too little food and water, thousands died even before the ships set sail from the islands, and the sea lanes to Hispaniola were strewn with floating corpses. In Hispaniola the Lucayans were thrown into the labor camps, where their death rate exceeded even that of the Tainos. This cruel practice raged on until there were no Lucayans left to enslave. Finally, a ship was sent out to search all the islands for people that might be hiding in caves or forests. After three years of diligent searching, only 11 people were found. When they were carried away, silence settled over the islands. It was almost as though the island people had never existed.

For the next 100 years, the Bahama Islands dropped out of history. The Spaniards, fired by their discovery of gold in Hispaniola, turned their attention to other lands—to Central America and the kingdoms of the Aztecs and the Incas farther south. There they found vast stores of this precious metal, and Spanish galleons laden with treasure sailed back to Spain. In order to use favorable westerly winds for the passage, they had to sail north past the belt of easterly trades, which brought them close to the Bahamas.

But the Spaniards had lost interest in these islands, and by 1640 the Bahamas remained unoccupied. The British and French made some attempts to take advantage of this vacuum, claiming certain of the islands as their possessions. Nothing much came of these gestures until a band of English settlers in Bermuda decided to seek a land where they could enjoy religious freedom—a land where ridicule or persecution of any person for his religious beliefs would be prohibited and justice would be guaranteed for all. The islands of the Bahamas, still virtually uninhabited, seemed to offer a favorable location for the realization of this dream. They renamed the islands the Eleutheras, from a Greek word meaning "freedom."

After several unsuccessful attempts, a large group of pioneers set sail from Bermuda in 1648 and steered for the long, narrow island that is still known as Eleuthera. Unfortunately, the colonists did not know that the northern end of this island and the small adjacent islands are protected from the ocean waves by a formation of large fringe reefs called the Devil's Backbone and that, even today, is difficult to navigate successfully. Somewhere close to shore their ship struck a reef and was wrecked. One man drowned and all the provisions aboard ship were lost. But the disaster occurred near a wide sandy beach, and the settlers managed to reach shore, bringing with them the little sailboat that had been designed for exploring shallow waters.

Near the beach there was a large cave where they found shelter, and for months they lived on "such fruits and wild creatures as the island afforded." But food sources were scarce on this windswept shore and starvation was an ever-present danger. Finally, the leader of the group set off in the little sailboat with a crew of seven to look for help. After nine days and nights, they arrived in Virginia, where local people responded generously to the story of need and sent a ship loaded with supplies.

In the next few years the pioneers explored the nearby islands and cays and established several settlements in

more favorable locations. Harbour Island is a small morsel of land, just 3 miles long and half a mile wide, but with a large, well-protected harbor.

The settlement on Harbour Island was later named Dunmore Town, after Lord Dunmore, one of the early British governors of the Bahamas. In the meantime, central control of the islands, tenuous as it was, had shifted to New Providence, an island south of Harbour Island. Nassau became the capital, and, officially, the Bahamas remained a British crown colony until 1973, when they were granted independence.

Signs of the historic past are still evident in Dunmore Town. The original settlers designed their houses with architectural features that they had known in their homelands. A style more suitable to a colder climate, with pitched roofs for shedding snow and dormer windows for economical space, was a special favorite. But the settlers made one important change, which brought the town into harmony with the tropical setting. Instead of the white and gray of Bermuda and New England, they added riotous color: yellow, like the golden Bahamian sunshine; all shades of blue, aqua, and lavender, like the incredibly beautiful waters that circled their shores; and red and pink, like the hibiscus and oleander that thrived in the warm climate and decorated the front yards and tiny back gardens. Night-blooming jasmine filled the soft night air with fragrance. In the center of the town, Temperance Square was awash with purple bougainvillea, covering an area as large as a city block with radiant color.

In the 1990s the population of Dunmore Town was about 1600. The majority are black. There is a white population descended from the original settlers as well as loyalists, who immigrated to the island during the American Revolution. Despite the racial mix, there is harmony between all the different factions (as we observed on Mauritius and Lombok), resulting in a community with a pleasant, peaceful way of life.

In recent years, people from the United States and Canada, anxious to escape the cold winters, have bought and remodeled a number of the old houses in town and built new houses outside the town. Half a dozen small hotels accommodate the vacationers who come to spend a week or two in this tropical Eden. Although tourism is an important factor in the economy, it is not large enough to seriously alter the character of the town. And, of course, in summertime the island returns to its unadulterated state—peaceful, slow-paced, smiling in the sun.

———

As one warm summer day follows another, the crystal clear waters surrounding the Bahamas, as well as the islands that lie in a long curving path just to the south— Haiti, Puerto Rico, the Virgin Islands—reach temperatures of 80 degrees or more. Under these circumstances the most violent storms can be born. The islands directly south of the Bahamas are especially prone to the development of hurricanes, because they are directly in the path of waves of low pressure caused by hot air rising over the Sahara Desert in the late summer. These waves are borne on the easterly trade winds from the coast of Africa to the Caribbean, where they encounter the hot humid air over the tropical seas.

A wave of low pressure in the upper atmosphere causes the air warmed by the ocean waters to be drawn upward, just as hot air is drawn up a chimney. The rising column of air is deflected by the Earth's rotation (the Coriolis effect), causing it to turn in a giant whirl as it rises. Thus, a tropical cyclone (a hurricane or typhoon) is born. The spiraling column of rising air is cooled to the point where it can no longer hold so much moisture, and clouds and raindrops form. This process releases heat energy, which warms the air again and reinforces the updraft. The winds increase in speed, more moisture-laden air rushes into the base of the spiral to equalize the low pressure caused by the updraft, and the updraft grows stronger as the water condenses out into torrents of rain. The spiral motion sends the raindrops spinning to the edge of the vortex, just as water swirling down a drainpipe is thrown to the periphery of the little whirlpool, leaving the center relatively empty.

A hurricane is like a thunderstorm multiplied many times and given a strong spiral spin. It strikes with a force equal to the simultaneous explosion of 1000 atomic bombs like the one that leveled Hiroshima. It can lift 2 billion tons of water a day, and this enormous load is emptied out of the storm as rain. It is literally true that the air picks up a piece of the sea and drops it again miles away.

Saturday, August 22, 1992, on Harbour Island was one of many idyllic Bahamian days. The sun was bright, and the many-hued sea of aqua and turquoise blue was reasonably calm, with only the usual waves breaking on the coral reefs and pink sand beaches. But for several days the radio and the television sets connected to satellite dishes were reporting the progress of a tropical storm—Andrew—which was moving northwest from

Puerto Rico on a track approximately paralleling the Bahama Islands.

During the afternoon and night of August 22, the winds in the storm picked up speed and within a 12-hour period, Andrew changed from a tropical storm to a full-sized hurricane, packing winds of more than 75 miles an hour. The path of the hurricane had also changed. It was now traveling straight west, heading directly at Dunmore Town, which lies at 25°30' north latitude. By noon of the next day, Hurricane Andrew was about 50 miles east of Harbour Island and was moving west at 15 miles an hour. It continued to gather strength. In a few hours it grew from a grade 1 to a grade 3 hurricane. This ominous news was received by television and radio in many of the houses in Dunmore Town—tiny, flimsy, wooden structures and sturdier concrete block houses (a favorite building technique used in recent years).

The fear created by this news is difficult to imagine. Most of the disasters that had struck the island had come with little or no warning—the raids of the Caribs, the slave ships of the Spaniards. But to know for hours that a major catastrophe is coming directly at you, and that there is no effective action you can take to protect yourself or your family—this is a terrible experience, dramatizing the utter helplessness of man in the face of nature.

No warnings were issued urging the local people to evacuate. There was no transportation available for a mass exodus and no place to go. At the North Eleuthera airport, several private planes were taking off but the commercial flights were cancelled. There were no shelters to provide any measure of protection for the population. Only a few of the houses had basements and those were very small. There was literally no place to hide.

Most of the individuals expected to ride out the storm in their own homes. All morning the people hurried to board up the windows, fasten the shutters, and say many prayers that by some miracle the storm would pass them by. No major hurricane had struck these islands for the past 30 years.

But every hour, the news grew more and more ominous. The eye of the storm was still headed directly for the town, and it was now characterized as a grade 4 or 5 hurricane, probably the most dangerous storm of the century.

At three o'clock in the afternoon the winds began to blow out of the north, intensifying rapidly, from about 40 miles an hour to 100, then 150. Heavy rain began to fall. A British naval vessel in the harbor clocked wind gusts of more than 200 miles an hour. Plywood panels that had been hastily nailed up were torn loose; windows were shattered. Rain and huge waves broke into many houses and flooded the streets that bordered on the harbor. The noise was almost unbearable. One witness said it felt like being trapped in a small room with several large jet engines open at full throttle. Others said it sounded like a hundred bombs going off in rapid succession.

Huddled in the corners of their little houses, every family tried to outwit the dangers and survive the storm. Reggie, the owner of a local taxi service, crouched in his house, watching the walls shake, the windows crumple. A crack directly overhead warned him of worse disaster. He ran out the front door and jumped into his van, which was parked outside, just as the roof of his house collapsed, obliterating everything inside. The van itself was picked up and rolled over by the wind, breaking several windows, but Reggie was unhurt.

Across the harbor and to the west, about 7 or 8 miles from Dunmore Town, several well-built vacation houses stood on a rise of land called the Ridge. The bishop of Rhode Island owned one of these houses, and he happened to be there on August 23. After boarding up the house as securely as possible, he retreated inside and for

3 hours watched helplessly as wind and water started to invade his home. About 6:00 P.M. the wind began to slacken and by 6:15 the eye of the storm had passed overhead.

Stepping outside, the bishop looked on a scene of wild destruction. Trees were down everywhere; the roof and door of his garage were a twisted mass on the ground. But the view of the sea was the most astonishing sight. As he described it:

> Normally we would have had about 4 feet of water at the steps going down into the ocean. However, the earlier wind from the north literally pushed the ocean out several hundred yards, and we had over a thousand feet of beach during the period when we were in the eye. Just before the eye had passed fully over us, a river of water, perhaps a hundred to 200 feet wide and several feet deep came rushing along the shoreline from the east at what must have been 50 miles an hour. Beyond this river lay a couple of hundred yards of sand between it and the ocean's edge, which was still several hundred yards from its normal place. Then the wind reversed itself and the ocean came flooding back in. It crashed against the southern edge of the river with explosive force— wild noise like sticks of dynamite going off and water spouts springing 40 to 50 feet into the air. Then it became a high wall, moving in slow motion toward the house and me.

This wall was later estimated to be 18 feet high. The bishop continued:

> It was at this point that I beat a hasty retreat into the house just as the first blasts of part 2 began to seriously build up on the south face of the house. I retreated quickly to the opposite corner of the house and crouched down in the corner of the

kitchen. Within a few minutes I began to notice the flexing along the north side of the roof beam. Although it was beginning to approach dark outside, I remember seeing daylight as the boards would pull away from the beam momentarily, and then it was as if an explosion had occurred under that section of the roof, and it simply sprang into the air and disappeared to the north. Shortly thereafter, I was able to see waves from the ocean washing over the sill from the patio and blowing across the living room floor. Then, several waves washed in about 18 inches high. Two other vivid memories are etched in my mind. The first is of the glass top on the dinette table being pushed by water from below and floating off the table. The second is of the refrigerator beginning to float across the kitchen in a haphazard fashion. At this point it seemed time to vacate the kitchen! I retreated once more, this time to the bedroom on the northeast corner of the house. From roughly 8 P.M. until about 10:30, I stayed in the bedroom "holding off the storm" literally by wedging myself against the door. I wedged my foot against the bottom and more or less stayed in that position for the next two hours. Around 10 P.M. the winds began to slacken, but I was afraid to venture out of the room until around 11:15 or 11:20.

In the morning of August 24, incredible scenes greeted the inhabitants of Harbour Island and the northern shore of Eleuthera. One man remarked that the land looked as though a giant had trampled over it, eaten up all the vegetation, and finding it unpalatable, had vomited it over the earth.

Two little villages, Spanish Wells and Current, were severely damaged by the surge wave (a tsunami created by enormous differences in barometric pressure). At Current, the wave funneled by the shoreline was 23 feet high. A customs official who had gone to the airport to supervise the departure of the last planes, went home to Current after the storm. He found that his house and all its contents had been swept out to sea—only a concrete slab remained. He had nothing left but the clothes he wore.

Several of the residents in this area had been in danger of being trapped in their homes. They saved themselves by sawing holes in the roof and escaping before the rising waters totally engulfed their houses.

The destruction to piers along the shores was very nearly complete. The wooden docks that had served Harbour Island were uprooted, their heavy posts hurled against the shore like battering rams. Lighter pieces were impaled on tree trunks and fences far from the water level. The main concrete pier in Dunmore Town was damaged but not destroyed. Many boats were broken up or sunk. All in all, the property damage was estimated to be several million dollars. Most of the inhabitants did not have insurance. They did, however, have something of much greater value. They belonged to a community that was accustomed to working together; the inhabitants all knew each other and in this moment of danger they helped each other. The homeless were taken in by friends or family, so no one was left without shelter. A few hours after the passage of the storm, every able-bodied person went out with a shovel or machete. (Power tools were always scarce on this island and that day there was no fuel or power to run them.) Slowly, methodically, they went about the job of cleaning up the island.

The response to the disaster in Dunmore Town was very different than it was in Florida, also struck by Hurricane Andrew. There was no flood of relief supplies, like those that inundated the distribution centers and roadsides in Homestead, Florida. Help did come but on a much more modest scale. Private individuals chartered planes to bring in food and water. The British navy supervised the distribution of these supplies. The U.S. Navy sent a 22-man detachment of Seabees to assist in the reconstruction. The Red Cross sent medical supplies and nurses. Money was provided by the Taiwan government, the Bahamian government, and private donations. The island of Jamaica sent supplies by ship.

These donations of aid from people who were not responsible for the Bahamians was reminiscent of the rescue mission sent to save the little band of ship-wrecked people on the beach near Harbour Island. The spontaneous outpouring of sympathy and help from many sources seemed to reinforce a sense of personal worth and responsibility. Although the need of individuals for food, water, and dry clothing was great, there was no looting in Dunmore Town. The stores remained stocked with supplies, and several of them were open just a day after Andrew had passed.

Destruction of property was extremely varied. Some buildings were completely wiped out. These were in general the flimsiest structures. Other buildings were almost unscathed, with perhaps a window knocked out or roof tiles blown away. Surprisingly, several little thatched gazebos at the edge of the bay were left virtually intact; only a little of the thatch was blown away. Hurricanes are notoriously capricious. Small variations in the configuration of the land and the seabed can make all the difference between destruction and survival.

To be directly in the eye of a hurricane is in some ways an advantage and in other ways a special hazard. Because the winds are calm in the eye, the total exposure to high winds lasts for less time than in those areas that are exposed throughout the entire storm to the intense winds that surround the eye. However, the

barometric pressure is lowest in the eye and is responsible for causing the fearful surge wave. Those regions that
lie on the northeast side of the eye, where the speed of
the hurricane winds is added to the speed of the forward
motion of the storm, receive the worst battering by the
wind. Small tornados are frequently embedded in the
wall of the eye, and there is some evidence that tornados
cut narrow swaths through Dunmore Town.

Hurricane Andrew's eye was about 11 miles in diameter—large enough to encompass all of Harbour Island.
The center passed just south of the town. In the late
afternoon of August 23, when Andrew struck Harbour
Island, it had reached its maximum intensity, with the
highest wind speeds and the lowest barometric pressure.
Amazingly, there were no deaths or serious injuries on
the island and only three deaths in the whole area.

When I visited the island six months after Andrew
had passed, I found that it had recovered to a remarkable extent. The process had been accelerated by the
diversity of the skills and resources of the population.
Painters, plumbers, mechanics, and carpenters were all
available. Local con-
tractors cooperated
in correcting the
most serious dam-
age first. Telephone
and electric service
was restored in the
town in the first few
weeks. Roofs were
rebuilt, starting with those
on the school, the clinic,
a rectory, and the two
major churches.

Then work began on restoring the landscape. Plants were imported to replace some of those that had been destroyed. Above all, nature cooperated in a spectacular way. The leaves returned to the denuded branches of the citrus and gumbo limbo trees. Soft green vegetation began to cover the bare spots, and an extraordinary profusion of flowers graced the island. Hibiscus and oleander were "blooming their heads off." To explain this unusual profusion, the local people cite an ancient belief that the salt spray that drenches the island during a hurricane is good for the land. There may be some truth in this. It is not the salt itself, because many plants are killed by exposure to salt, but ocean water carries many important trace elements. Normally, the soil of the Bahamas is deficient in several of these, notably manganese and zinc. So, when Andrew sprayed the island with seawater, then rinsed it with drenching rains that washed away most of the salt, it may have left a residue of essential elements, causing a spectacular burgeoning of new life.

A comparable resurgence of spirit took place in the people of Harbour Island. Unlike the experience in southern Florida, where there was much trauma as a result of Hurricane Andrew, in Dunmore Town there was a sense of pride and accomplishment. For months afterward, whenever the hurricane was mentioned, faces lighted up. Everyone was eager to tell his or her own tale of terror, of survival, and triumph over the disaster. As the English historian Arnold Toynbee so wisely recognized, a severe challenge successfully met acts as a stepping stone to higher levels of achievement.

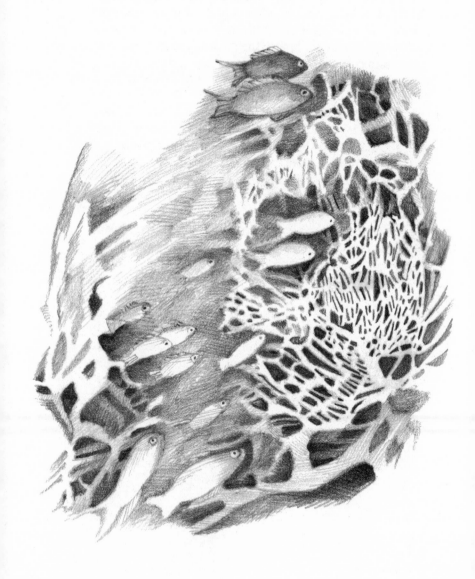

Chapter 12

CORAL ATOLLS

Living Islands

Full fathom five thy father lies;
Of his bones are coral made:
Those are pearls that were his eyes:
Nothing of him that doth fade,
But doth suffer a sea-change
Into something rich and strange.

—SHAKESPEARE

There is no more perfect example of the constructive artistry of nature than the making of a coral atoll. These tiny gemlike islands are not formed in a flash, as Surtsey was born from the sea, but they form slowly, step by step in a process that may take 50 or 60 million years. Grain by grain, polyp by polyp, a fringing reef is constructed in the warm embrace of the tropical sea. And as it grows, it protects and nurtures its own little piece of that sea—a beautiful lagoon.

In the century before Charles Darwin's voyage on the *Beagle*, world travelers brought back descriptions of the rings of living coral that they had seen in the Pacific Ocean. How could these mysterious formations be explained? It was observed that reefs exist only in shallow water, less than 150 feet deep. But the Pacific Ocean on the whole is immensely deep, as much 30,000 feet, and still it is speckled with thousands of tiny islands and reefs. So, what was all the coral growing on and how were the reefs established in the first place?

Several scientists attempted explanations. Charles Lyell, a highly respected geologist, suggested that these "lagoon islands" were made by coral growth forming on submarine volcanoes along the rims of their craters.

Starting with this hypothesis, Darwin created a more comprehensive theory that explained many phenomena previously thought to be unrelated. He suggested that the process begins with violence as a volcano erupts in the depths of the sea and raises its fiery head above the ocean surface. When the hot gases and the pressure that drove the volcano up are dissipated, the mass of the new mountain slowly subsides and settles back toward the seafloor. Rains erode it, taking part of its substance into the sea. Gradually, the land above water level is reduced and shallow waters surround the island. The presence of this shelf just beneath the surface triggers the formation of coral, especially around the perimeter where the conditions are very favorable for reef building. Year after year, the fringing reef continues to grow, keeping pace with the subsidence of the land on which it rests. Eventually, the mountain sinks beneath the ocean surface, leaving a calm, shallow, pale blue lagoon surrounded by a necklace of coral and beyond that the dark, wave-girt sea.

Other scientists, unwilling to accept the concept of subsidence, proposed different theories. For example, the ocean floor, they maintained, contains many irregularities in surface—banks of sediment that could have served as platforms for coral reefs, especially at times of low sea levels. These reefs grew until they reached the ocean surface. As the water levels continued to drop, the tops of these formations approached sea level, and the corals grew upward, making fringing reefs.

During the past century, it has been recognized that tremendous changes in sea level have occurred—such as those that accompanied the end of the last ice age—and it has also been discovered that the floor of the Pacific Ocean is dotted with hundreds of guyots, flat-topped seamounts that rise thousands of feet above the ocean floor. These are volcanic islands that were once above sea level, as evidenced by dredgings of shallow marine fossils from their tops. When sea levels rose, the peaks of the islands were planed off by wave action. Coral reefs may have been established on them before they became entirely submerged, but in many cases they could not keep pace with crustal subsidence, and the coral did not survive in the cold, dark depths of the sea.

The arguments supporting these different explanations went on until improved technology revealed more information about the foundations on which coral atolls are resting today, the way coral reefs form, and how they grow and die.

The active portion of a coral reef is a thin fringe covering the surface of the bony structure. There thousands of tiny animals feed and grow, secreting the limestone base that gradually builds the skeleton of the reef. These coral polyps may be golden yellow, magenta, or vibrant blue or green. Coral grows in many different forms that frequently resemble various types of plants on land—a bunch of lettuce, a mushroom, a cauliflower, or a bush in bloom. And for a long time scientists assumed that coral was an aquatic plant. Then, the surprising discovery was made that the coral is made up of polyps—simple animals containing just two layers of cells.

Because a polyp is an animal, it must obtain its energy by the ingestion of organic molecules in which the sun's energy has been stored by plants using the process of photosynthesis. Tiny algae named zooxanthellae live with the coral polyps in a close partnership. The algae capture the energy of the sun and synthesize organic compounds that provide an ideal food for the coral. The coral polyps, on the other hand, provide a sheltered environment, and their waste serves as nourishment for the algae. This symbiotic relationship, beneficial to both partners, has been so successful that it is characteristic of most coral formations around the world. But in order to raise a good crop of algae, the living portion of the coral reef must be near the ocean surface in order for the zooxanthellae to receive a considerable amount of sunshine.

The coral polyp builds its bony skeleton from the calcium it extracts from the sea. Most of this activity takes place during daylight hours, using the energy of sunlight. The growth of the reef waxes and wanes in a regular rhythm day by day and year by year, producing a pattern of rings in the stone—a beautifully detailed record of the sunlit hours on earth. Incredible as it may seem, it is possible to read the fine print of the record back as far as 400 million years ago, when coral reefs first appeared on

Earth. This record tells a fascinating story about the diurnal cycles and the changes throughout time. The oldest reefs have 400 diurnal growth lines in each annual band, while somewhat younger corals (270 million years old) have 387 lines per year. Modern ones have about 360, suggesting a progressive decrease in the number of days in a year since the earliest reefs were formed.

The little coral polyp reproduces itself in several ways. New life most frequently takes place by budding. A miniature new polyp forms—an identical copy of its parent—and remains attached to the parent as it grows. Eventually, it also buds, and in this way a colony of polyps develops. This cluster of organisms is an association for which there is no parallel in human life. More closely related than a family, the polyps remain attached to each other and have identical genetic endowments— a cluster of clones that acts like a single individual. As budding proceeds, the individual polyps line up side by side, building a bony structure characteristic of the particular species to which they belong. In some, the units are very tightly spaced. In others, each separate cup is surrounded by a broad foundation of limestone.

There are hundreds of coral types, varying all the way from the densely packed whorls of brain coral to the open shapes of staghorn coral. Under given conditions of wind and wave and sunshine, certain shapes are more favorable than others. In the rough waters of the open ocean, the coral heads that survive the longest are massive ones with compact shapes that resist the pounding of the sea. In quiet, protected waters, the delicate, branching forms offer more surface for intercepting the light and sifting plankton from the water. The individual coral polyps and their algae are working units in a larger and more complete structure that is advantageous to them all.

In addition to asexual budding, the coral reproduces sexually, and each polyp has the enviable ability to play either the male or the female role, switching sex apparently at whim. At certain seasons of the year, sperm or eggs develop within a fold of internal skin. The sperm are released when they are ripe and drift until they are drawn into the mouth of a polyp containing mature eggs. The fertilized eggs soon produce tiny larvae covered with microscopic hairs. When these are released, they use the little hairs like arms to swim and guide themselves through the water. Amorphous as tiny clouds, they take on an almost infinite number of forms. Drop-shaped one moment, they may be sausage- or crescent-shaped the next. Millions of these larvae—or planulae—emerge from a living reef at certain times of the year.

Almost all of the planulae are eaten or die before they find a place where they can attach themselves and start to grow. They are bare and totally defenseless in a predatory world. Those that survive must encounter a very special set of circumstances before they can start a new colony. If they settle down on ordinary ocean bottom, they will be buried by shifting silt and sediment. In some mysterious way, they seem to "know" that they must find a hard surface to which to attach themselves—a surface raised above the bottom of the sea. A submerged wreck or an empty portion of an old reef is the foundation they must use in order to begin their work of fashioning a new coral community.

To the casual observer the coral community seems to be a dazzling assembly of many beautiful individual things, all competing for food, space, and sunlight. Actually, this galaxy of creatures has evolved intimate, closely interdependent relationships. As the new coral community divides and multiplies, it builds an underwater castle with many rooms and corridors providing shelter for its symbiotic algae and safe hiding places for the multitude of fish that are attracted to the growing reef. Crinoids, sponges, sea worms, and anemones take up residence in the sheltering, richly varied environment. The reef, including all the species that inhabit it, functions as one giant living unit—a superorganism.

The coral skeleton provides the "bones," or the supporting tissue; its porous construction with many caves and tunnels acts as a circulatory system, allowing the lifeblood of the sea to reach every "cell" of the body. The efficiency of the system is enhanced by the presence of sponges and many other organisms that burrow into the reef, creating an even more elaborate network of passageways and pumping water through the pores of their own bodies. During the day, vast numbers of fish hover in dense schools over the reef, feeding on plankton and microscopic animals that float in the sea. The rain of fecal pellets dropped by the fish is consumed by the coral polyps. These contain phosphates and nitrates, which are excreted in turn by the polyps and passed along to the zooxanthellae, providing essential fertilizer for the coral's vegetable garden. Besides producing organic nutrients for the coral polyps, the algae also manufacture a lime deposit that helps to cement loose sand and sediment and repair minor damages to the skeleton.

Sea urchins also perform an important maintenance function. They clean infected portions of the reef, feeding on the blue-green algae that grow on torn coral tissue and that might threaten the health of the reef. The coral skeleton is constantly trimmed and shaped up by butterfly fish and bristlecone worms. Excessive growth of any one part of the reef is kept in check by a kind of chemical pruning. Inside each polyp are filaments that digest animal tissue, and when these protrude through holes in the body wall, they can

absorb the polyps of neighboring corals, preventing encroachment.

The reef has a communication network like a giant nervous system. Chemical signals carrying vital information about sex, food supply, injury, and impending danger are discharged into the water and circulated by the sea. As the subunits of the organism receive these messages, they are stimulated to respond in ways that coordinate the feeding, reproduction, growth, and regeneration of the larger whole.

Thus, the multitude of "individuals" that have lived inside the reef have all worked together throughout millions of years to produce one giant unit that breathes and grows and reproduces itself—a living island.

One of the most significant biological discoveries in recent years is the importance of integrated systems in nature. This insight came as a surprise to modern man. The principle of the survival of the fittest, emphasizing competition, the struggle for dominance, and survival at the expense of others, has been so successful in explaining the evolution of life on Earth that evidence of cooperative behavior and symbiotic relationships have been seen as odd exceptions to the rule. Now, we are beginning to understand that these integrating activities are essential to the way nature constructs ever more complicated beings by putting together simpler things. These two forces—competition and cooperative behavior—are equally important. The tension between them provides a balance that allows constructive change, growth of variety, and increasing complexity as time goes by.

Some of our knowledge about coral reefs and the formation of atolls came about as a result of the development of the atomic bomb. After World War II had been decisively terminated by the use of this terrifying new weapon in Hiroshima and Nagasaki, the United States Navy decided to assess its destructive powers when used against ships at sea. A tiny atoll named Bikini at the northern edge of the Marshall Islands was selected as the site of this experiment.

In preparation for the test, the atoll was drilled and studied by seismic refraction. The measurements showed that about 4000 feet of coral rested on a basement of basalt rock, volcanic formations like most of the ocean floor. These facts were compatible with Darwin's theory of subsidence. The central volcanic peak sank beneath the ocean level millions of years ago. The growth of the fringing reef kept pace with the rate of subsidence, and eventually 26 little islets formed around the perimeter of the reef.

Bikini is the largest of these islands, lying on the windward side of the lagoon. It is a pretty atoll with many coconut palms and pandanus trees and white sand beaches. The lagoon is very large—blue-green and as translucent as an aquamarine.

One hundred and sixty-seven people lived in this tranquil tropical environment, maintaining a gentle and unsophisticated lifestyle. They fished in the spacious lagoon, hunted turtle eggs, and picked the breadfruit and coconuts that grew near their village. They were skilled builders of outrigger canoes for exploring the nearby islands and expert spear fishermen, catching the delicious fish that inhabited those quiet waters. A peaceful and trusting people, they had been converted to Christianity in the early years of the twentieth century, and they were especially open to religious appeals.

One Sunday in June 1946, Commodore Ben Wyatt of the U.S. Navy met with the people after church services.

He explained that their island was needed for a project that would benefit all mankind. He implied that an authority higher than any on Earth would be pleased if they decided to cooperate by vacating their homes and moving temporarily to another island.

After consultation with the others, the chieftain Juda replied to the commodore, saying, "If the United States government and the scientists of the world want to use our island and atoll for furthering development, which with God's blessing will result in kindness and benefit to all mankind, my people will be pleased to go elsewhere."

All 167 members of the community were transported to a nearby uninhabited atoll, Rongerik, taking with them the thatch from their 26 houses, the dismantled church, and community hall. They believed the assurances that this was a temporary move and that they would soon be brought back to Bikini. Unfortunately, this was just the beginning of a woeful saga. They became nomads of the atomic age and were repeatedly relocated to other Pacific islands, where they found only unhappiness.

After the inhabitants had been removed, the U.S. Navy assembled an impressive collection of ships to serve as targets for the atom bomb explosions. There were 80 obsolete vessels: battleships, cruisers, submarines, and landing craft. Several Japanese and German war prizes were there, including the infamous battleship *Nagato*, the flagship of the Japanese admiral who directed the attack on Pearl Harbor, as well as a number of U.S. ships of World Wars I and II, including the *Arkansas* and the *Saratoga*. These target ships were moored in the calm, beautiful lagoon, and many other vessels were assembled nearby to observe the effects of the blast.

On July 1, 1946, an atom bomb was exploded in midair over Bikini. Five large ships were sunk immediately. Three weeks later another bomb was detonated underwater with even more dramatic results. A 43-foot wave was created. Two million tons of water and sediment were hurled more than a mile upward and then fell on the ships. The *Saratoga, Arkansas, Nagato*, two submarines, and several other ships were mangled and submerged. In all, 12 of the large target vessels were drowned in these first two explosions, but more explosions were still to come.

Between 1946 and 1958, 23 atomic tests were performed at the Bikini lagoon. The most violent one occurred in 1954. A hydrogen bomb 1000 times stronger than the one dropped on Hiroshima was detonated on the lagoon's northwest side. Named Bravo, this was the most powerful bomb ever exploded by the United States. It has been said that the combined power of all the weapons fired in all the wars of history would fall short of that released by Bravo over the Bikini lagoon. The navy had calculated that if a southerly wind prevailed at the time of the experiment, it would blow the radioactive cloud and debris into the uninhabited sea to the north. The morning of March 1 dawned with a wind blowing toward the northeast. Although this was not ideal, the northeast direction was deemed safe enough, and the bomb was exploded right on schedule.

Almost immediately it became apparent that a serious mistake had been made. The wind shifted again and was blowing toward the east. In the hours that followed, radioactive dust fell on the naval vessels that were observing the shot from a position southeast of the atoll and on the islands of Rongelap, Rongerik, and Utirik, as well as Bikini itself. The fallout also drifted over a Japanese fishing boat, ironically named the *Fortunate Dragon*, which happened to be northeast of Bikini and had not been noticed by the patrolling aircraft.

It is not known precisely what dosage of radioactivity the 23 fishermen received, but the best estimate is 200 roentgens. One death and a number of illnesses resulted from this strong exposure. The nearby atoll of Rongelap also received very high radioactive fallout, and it was only by a very lucky chance that the natives were saved from a lethal dose. If their island had been 30 miles farther north, they would have received a dosage that would have meant certain death. The atoll of Rongerik was also swept by the radioactive plume, but the Bikinians had already been relocated, because they were starving on that little atoll where the fish had become poisonous and coconuts were scarce.

The ships that had remained afloat in the Bikini lagoon seethed with radiation. After initial decontamination efforts failed, they were towed 200 miles to another atoll—Kwajalein—for further cleanup measures. When these also failed, the derelicts were sunk there in target practice from Hawaii and the U.S. West Coast. In the years following 1954, Kwajalein was regularly used as the target for missile practice. The technique was so accurate that missiles fired from California, 4800 miles away, seldom missed their target. They slammed into the lagoon or the reef of Kwajalein.

In the meantime, the Bikinians had been relocated again to the island of Kili, about 200 miles farther south. Kili was much smaller than Bikini, and it had no shallow lagoon or sheltered harbor. On these unprotected shores, battered by wind and wave, fishing was very difficult. Canoes were useless, and the skills that the Bikinians had developed over the generations were lost. Hunger was an ever-present reality. These people, who had once been proud and self-sufficient, became wards of the U.S. government.

The waters of Bikini lagoon are tranquil now, concealing a ghost fleet of 21 vessels. The *Saratoga* lies on the bottom, its flight deck only 100 feet below the surface. It settled upright on its keel and planes are still arrayed on the hangar deck.

But the soil on Bikini was so contaminated that the inhabitants were not allowed to return there until many years later and then only briefly. Cleanup efforts began in the late 1960s, and in the early 1970s a resettlement of the Bikinians began, but in 1978 it was decided that the island was still too heavily impregnated with atomic bomb debris. The food grown on the island was radioactive. In the bodies of the inhabitants strontium 90 and cesium 137 had reached dangerous levels. Cesium 137 had penetrated the soil and the water supply. Although this element does very gradually decay, nearly 100 years more would be required before the radiation reached acceptable levels. The Bikinians were again exiled to the island of Kili.

As late as 1992 there was talk of decontaminating the island of Bikini by scraping 12 inches of topsoil from its 560 acres, but this option was considered too expensive, and no one knew how to dispose of the radioactive soil. Today, half a century after the inhabitants of Bikini agreed to cooperate on a "project that would result in kindness and benefit to all mankind," they are still wanderers in a world turned upside down.

Eniwetok, another coral atoll in the Marshall Islands, was also used as an official U.S. testing ground for atomic weapons. Its fringing reef encloses a large lagoon, circular in shape with many tiny islets, set like baguette diamonds around the ring. During World War II, Eniwetok was captured from the Japanese by U.S. forces, and its fine anchorage was made into a naval base. After it was designated a testing site, the base was temporarily abandoned and, like Bikini, its inhabitants were evacuated to other atolls. Also very similar to Bikini, the drilling at Eniwetok in 1952 revealed a very old volcanic base at a depth of about 3000 feet.

Many atomic weapons were tested there, including one hydrogen bomb, which vaporized the tiny islet on which it was detonated. The last test was made on Eniwetok in 1956. Later, 111,000 cubic yards of contaminated topsoil were removed and buried in a huge concrete burial mound on the islet of Runit. In 1980 the people of Eniwetok were given an opportunity to return. Their first crops, however, were found to be too contaminated, and, as at Bikini, the people were evacuated again.

⟶

Other nations followed the lead of the United States in building and testing atomic weapons. The French owned several atolls in the Pacific Ocean that could be used for this purpose. The first chosen was Mururoa, which lies at the southeastern tip of the Tuamotu Archipelago in French Polynesia, 700 miles southeast of Tahiti. In 1966 a midair nuclear bomb was exploded and caused some concern, because the southeasterly trade winds that sweep across French Polynesia could possibly carry some radioactive fallout to Tahiti, a densely populated and very important tourist site. After 1975 the tests were conducted underground, but international concern was expressed about the underground explosions when it became known that the coral rock of the atoll had been badly fractured. In fact, the 29 nuclear weapons France exploded there between 1975 and 1980 were said to leave the atoll "looking like a Swiss cheese."

A running battle between environmental groups and the French government continued for many years. In July 1985 the *Rainbow Warrior*, a ship owned and operated by the environmental organization Greenpeace, was being prepared to lead a protest at Mururoa. Before it could take place, the ship was attacked and sunk in the Auckland Harbor in New Zealand, drowning the photographer Fernando Pereira. This attack embarrassed France and gained international attention for Greenpeace.

In 1992 France declared a moratorium on nuclear testing. In 1995, however, it announced plans to set off eight nuclear blasts between September and May at Mururoa and Fangataufa, the neighboring atoll. The *Rainbow Warrior II* set sail with two companion vessels, hoping to reach Mururoa on the tenth anniversary of the 1985 attack, but the French navy was waiting for them. Commandos stormed the Greenpeace ship on July 10, thwarting any attempt to land protesters on the atoll. Some 150 men in black helmets and jumpsuits boarded the ship at dawn, knocking out doors, smashing windows, and taking two dozen people captive. About ten of these were held for questioning and were allowed to return to the ship only if they promised to leave Mururoa within hours.

In the meantime, Greenpeace scientists had been conducting chemical analyses in the area and had found radioactive isotopes in samples of plankton 12 miles from the atom bomb sites. These measurements confirmed results from sampling done in 1987 by Jacques Cousteau. This was a significant finding, because plankton is at the base of the oceanic food chain. Many fish and even whales depend on plankton for their nourishment. The dangerous isotopes are preserved in their body tissues and may be carried hundreds of miles away, where they serve as a food source for island populations.

⸺

These activities have not exhausted the experimental and disposal needs of modern man. As long as warfare remains an important tool in the pursuit of power, other types of lethal weapons will be tested and must eventually be destroyed. The little coral atolls offer a tempting solution for siting these dangerous activities. They are relatively defenseless and "not good for much else."

The United States owns a small atoll named Johnston in the central Pacific Ocean. It is 715 miles southwest of Honolulu. Small and comparatively flat, it rises only 44 feet at the highest peak. It is a prime example of "not good for much else." There are no sources of fresh water, rainfall is sparse, and vegetation is limited to bunch grass and herbs, but it was once frequented by many birds, and guano deposits were briefly a source of revenue. These deposits were exhausted by 1908. Twenty years later, the atoll was declared a bird sanctuary. After World War II the atoll served as a U.S. Air Force base, and it was used during the 1950s and 1960s for nuclear testing. One end of the atoll still remains radioactive.

Later, Johnston was considered as a test site for biological weapons, but this possibility was rejected, because it was recognized that the large population of migratory birds might carry their lethal infections to Hawaii or back to the continental United States. Beginning in 1971, Johnston was used as a collection point for thousands of tons of mustard and nerve gases, and a facility for incinerating these weapons went into operation on the island in the early 1990s. Dioxin and furan are sometimes released into the air, and all personnel on the island are required to have their gas masks ready. In 1998 it was discovered that the containers in which the chemicals are stored are leaking.

Although the northeasterly trade winds do not favor the drifting of plumes from the incinerators (or the flights of birds) from Johnston Atoll to the Hawaiian Islands, there are seasonal changes in prevailing winds and sometimes tropical storms. Could a small wind-shift cause another—and more widespread—human tragedy?

These attempts to harness the most powerful and unforgiving forces yet discovered in nature draw attention to the large margin of error in human endeavors.

Even with all our knowledge and our astonishing computer technology, chance and miscalculation still play important roles in the outcome: the undetected presence of a fishing boat; a small shift in wind direction; and the ability of the soil, vegetation, and life in the sea to hold and transmit the lethal rays released by atomic forces. These are small intimations of the terrifying powers that could be set loose by an unforeseen event, a tiny error in our best laid plans that would nevertheless devastate the human environment.

Ironically, the reefs at Bikini and Eniwetok are recovering, demonstrating the miraculous ability of the living reef organism to heal itself. In the lagoon at Bikini, the 21 sunken vessels have been decorated with a shroud of coral growth. Even the wounds from the direct hits with powerful thermonuclear explosions are slowly filling again with healthy coral growth. The billions of tiny living things are all cooperating to rebuild and mend the damage wrought by man. In fact, coral is the most successful form of life that has inhabited the Earth. It is the oldest and the largest and the most enduring.

There is another atoll, not far from Kwajalein and Bikini, where nature has conducted a different kind of experiment, testing the minimum conditions for the successful propagation of life. At Pingelap, three tiny islets form a broken ring enclosing a central lagoon about a mile and a half in diameter. The land is barely above sea level—nowhere more than 10 feet—but it has had human habitation for 1000 years, and at one time it had a population of nearly 1000 people. It is not known where the original settlers came from, but they brought with them a system of government ruled by hereditary kings.

In 1775 the powerful typhoon Lengkieki struck the island, driving storm surge waves and wild winds over the island and wiping out 90 percent of the inhabitants. Many of the survivors died a lingering death from starvation, because all the vegetation, including the coconut palms, breadfruit, and banana trees, had been destroyed. The islanders had nothing to sustain themselves but fish.

It is estimated that within a few weeks after the typhoon there were only 20 or so survivors, including the king and a few members of his royal household. Reproduction in this small, isolated population led rapidly to increased numbers, but after several decades it became evident that certain previously rare genetic traits were appearing with greater frequency.

—

The blueprints for every organism, including mankind, are carried by genes, which are present in pairs in each cell of the body (except the ovum and sperm). In each of these pairs, one gene comes from the mother and one from the father. Together, they control a single physical characteristic, such as the color of the eyes or the texture of the hair. Usually, one of the genes is more effective than the other in producing a certain trait. It is said to be dominant and the other one is recessive. Brown eyes, for example, are usually dominant over blue.

In the normal course of affairs, the shuffling of the genes from one generation to another creates the variety of human beings we see in the world about us. But once in while a gene undergoes a change, perhaps spontaneously or perhaps by the impact of an X ray or a cosmic ray. The result is a mutation, which alters the blueprint carried by that gene. In the vast majority of cases, the mutated gene would be deleterious to the organism, but fortunately it is usually recessive, and the dominant gene controls that particular trait. However, the recessive gene is passed down to children and grandchildren just like any normal gene. Those individuals who possess a defective gene that is not apparent, but which can still be passed on, are called carriers. If two such individuals marry, the resulting offspring has a pair of mutant genes, and the result may be expressed as a disability that undermines the health and well-being of this individual. The smaller the breeding population, the greater the chance that two carriers will marry, thus duplicating undesirable genes.

In Pingelap, with only 20 or so inhabitants, the odds of such combinations occurring was very large. In the fourth generation after the typhoon, a new eye "disease" began to show up. Infants with this disability appeared normal at birth, but at two or three months they began to show an extreme sensitivity to light. When they became toddlers, it became apparent that they could not see fine detail or small objects at a distance. At four or five years old, they could not distinguish colors. The name *maskun* (not-see) was coined to describe this condition. It occurred with equal frequency in both boys and girls, otherwise normal, bright, and active children.

Oliver Sacks, a neurologist who visited Pingelap in 1993 and studied the incidence of this eye abnormality among the inhabitants, reported: "Today, over 200 years after the typhoon, a third of the population are carriers of the gene for maskun, and out of some 700 islanders, 57 are achromats [color-blind]. Elsewhere in the world, the incidence of achromatopsia is less than one in 30,000—here on Pingelap it is one in 12."

In cases of color blindness, the cells of the eye that are designed to perceive color (called cones) are affected.

A mild form of color blindness is relatively common in the population at large. It involves just certain of the cones, say those that distinguish between red and green or between blue and yellow. But in more serious cases, all the cones are dysfunctional. The ability to see fine detail as well as color is affected. The eye is much more sensitive to light and is blinded by bright sunlight.

In Pingelap most of those born with maskun never learn to read, because they cannot see the teacher's writing on the blackboard or the print in the schoolbooks unless these are held very close—about 3 inches—from the eye. These children cannot enter into rough-and-tumble outdoor sports. As adults they cannot work outdoors in strong sunlight. Unusual frequency of color blindness has been identified on other islands with small populations—on Pohnpei, an atoll near Pingelap, and on Fuur, a little island off the coast of Denmark.

There are other disabilities that result from inbreeding, for example, increased frequency of cleft palate and harelip. I have witnessed this in the small isolated community of Spanish Wells in the Bahamas. On the lonely island of Tristan da Cunha in the south Atlantic, halfway between Africa and South America, half the population of a few hundred people suffer from symptoms of chronic asthma. The island was settled by a small group of British sailors in 1817. Geneticists believe that at least one of the founding fathers must have carried an asthma-susceptibility gene, which was passed down through the population.

These cases illustrate one of the most important reasons why very small populations—containing fewer than 50 members—may be on the road to extinction.

The tales of these tiny atolls where strange experiments have been conducted both by human beings and by nature stand in stark contrast to the pristine physical perfection of a coral island. Bora Bora is a resplendent example. Seen from the air, it looks like an exotic butterfly that has alighted on the surface of the sea. The outspread wings are an iridescent blue and in the center a small dark body sparkles in the sunshine.

Bora Bora is an almost atoll. Not yet totally submerged, its twin volcanic peaks still stand tall. The shining leaves of palm trees cascade down the mountainsides and spill over onto white coral sand beaches that circle the shores. On the outside edge of the fringing reef, pounding waves throw up fountains of spray. But inside the reef the water is still and a clear transparent aqua, like Venetian glass.

Bora Bora and the other islands of this archipelago were settled by Polynesians from the Marquesas about the ninth century A.D. When Captain Cook visited Bora Bora on his first voyage in 1777, he found a small community of warlike people, governed by a powerful chieftain. They were much feared by the inhabitants of the neighboring islands. Later, the island became a royal retreat and one of the most sacred shrines in the islands.

The Society Islands, of which Bora Bora is one, were annexed by France in the late nineteenth century. During World War II, Bora Bora had a U.S. naval base, and one of the small motus (islets) on the reef was a U.S. airbase. Fortunately, this island was not used as a target for missile practice or as a test site for atom bombs. Instead, its exotic beauty inspired James Michener's description of Bali Hai in his *Tales of the South Pacific.*

I first visited Bora Bora in 1973, and it was there that I learned to snorkel. I had not tried this sport before; the idea of being confined within a mask had never appealed to me. But when I arrived at Bora Bora,

I discovered that it was known to be one of the most wonderful places in the world to view the underwater life. People had come from near and far—Australia, Italy, Canada—bringing their scuba and snorkel equipment with them. The conversation every evening was about the remarkable sights that had been seen beneath the shining face of the sea.

One day I borrowed snorkel equipment from the hotel and walked down to the shore. Standing in shallow water up to my waist, I put on the mask and leaned over far enough so it broke the mirrored surface. Then, like Alice, I was suddenly transported into a wonderland.

A school of tiny electric blue fish swam just beyond my reach. They turned in perfect synchronization, reflecting the sunlight in flashes like blue lightning. In an instant they were gone. Just beyond where I stood—so near I might have touched it—was a cluster of coral entirely covered with tiny tube worms. Extending their pale pink plumes, they moved gracefully back and forth searching the water for food. Two sea urchins waved their forests of spines, and on an empty conch shell a gorgonian, looking for all the world like a bush in full bloom, trailed gold fronds in the wave surge that stirred the waters like a heartbeat.

Lured on by one wonder after another, I swam slowly out from shore and in a few moments I was floating over an underwater garden. The sea was only about 6 feet deep and very clear. The sun shone through with undiluted brilliance. Many types of coral were growing there—daisy coral like tiny blossoms, gold leaf and yellow cauliflower coral, and staghorn coral tipped with cobalt blue. Brilliantly colored fish moved among the fronds and branches. Drifts of pale transparent fish hovered near a mound of gold leaf coral and darted back within its recesses as I approached. Beautiful iridescent angelfish moved like butterflies from flower to flower.

An orange clownfish peered cautiously out between the waving arms of a pale sea anemone. This clever little fish manufactures its own protective coating and thus can live within the circle of the poisonous tentacles of anemones without suffering any ill effects and is protected from all its enemies. As the clownfish rubs itself lovingly against the anemone's tentacles, a thin layer of mucus is produced, a chemical signal to the anemone saying: "This is me—the clownfish—don't sting me. I'm your friend. I will swim just outside your fronds and lure bigger fish in near where you can catch them easily." The gaudy color and striking markings of the clownfish make it a very effective decoy.

From that day on, many of my waking hours were spent in a mask and snorkel, and at night my dreams were filled with vivid images of the life beneath the sea. One of the most striking impressions that has remained with me was the view of the inside of the fringing reef where hundreds of sea fans—purple and pink, lavender and gold—moved together in response to the ripples stirred by winds in the lagoon. It was a breathtaking sight, like a corps of ballerinas waving their graceful arms. These soft-bodied gorgonians find a coral lagoon an ideal platform on which to erect their delicate fans where the sun shines through and the rough pounding of unprotected sea is broken and moderated.

In these castles beneath the sea there are many rooms to explore. In every crevasse there are new sights to discover and then another and another. One day I had become separated from my "buddies," and I entered an enclosed coral room. The walls that surrounded me were finely constructed. Like leaves and branches and pinnacles, the coral made a filigree of exquisite detailing that would have shamed the builders of the Taj Mahal.

I was so absorbed by this sight that I did not notice I was not alone in this enclosure. A large barracuda was sharing it with me. At least 7 feet long and slender as a torpedo, the barracuda's body was the color of metallic silver. Its underslung jaw was partly open, revealing a row of razor-sharp teeth, and its shoe-button eyes were focused unblinkingly on me.

Many thoughts tumbled through my head—scraps of information and cautionary advice I had heard over the years. If you see a barracuda, swim slowly away from him—never toward him—and make no sudden gesture, because if it feels threatened it will attack. But I could not swim away. I was trapped in this exquisite coral room, and the barracuda was occupying the only exit— the doorway to the enclosure. Desperately, I studied the walls. Perhaps there was another way out. Perhaps I could swim over one of the walls, although I knew immediately that there was not enough clearance for safe passage, and if I tried I would be impaled on the sharp branches of the staghorn coral. The stern, staring eyes of the barracuda twitched as he took in my odd appearance. Treading water in my orange wet suit, I must have looked like a giant clownfish caught away from its anemone.

Suddenly the sinuous body moved, reflecting a flash of sunlight. As he swam away from the doorway, I moved slowly toward it. And, keeping as great a distance as I could between the barracuda and me, I reached the exit and the relative safety of the open sea.

—

The barracuda disappeared into the dim recesses of its underwater world, and I returned to the world of man. But the vivid memories of the life beneath the sea remained with me. Sometimes, as I walk the concrete streets of our cities and ride the elevators up the towers of glass and steel, I think about the many-chambered and multistoried reef. We human beings, too, have built intricate and marvelous dwelling places, and although these may seem hard and enduring, they are more easily destroyed. A hydrogen bomb could pulverize these structures, which have no power to heal themselves.

Four hundred million years of evolutionary progress lie between the invention of the coral reefs and the cities of man. Human beings have progressed far in those millennia. The mind has evolved, and language has been discovered. Music and art have burgeoned, and many secrets of nature have been unlocked. But while all this has been happening, we have failed to understand and develop the remarkable power of forming communities with other living things, all working together for the good of the whole. We have not done this even with our own species. Self-defense and the predatory drives are still dominant in mankind, as they are in the barracuda. These drives—brutal and elemental—lead us as we invent more and more lethal weapons of mass destruction. If we continue down this path we will reach the end of our spectacular evolutionary progress. The Earth itself and all its beautiful coral reefs will surely survive but we will not.

Chapter 13

ATLANTIS

The Enduring Myth

It really isn't anywhere!
It's somewhere else instead.

—A. A. MILNE

Mysteries are the fabric that dreams are made of. They whisper of wonderful secrets that are hidden from us—secrets that may confer a special beauty or significance to our existence. We have been surrounded by mysteries all of our lives. The beautiful blue sky and the clouds conceal from us every day the magnificent universe of stars and galaxies emitting energies that we do not understand. Three-fourths of our planet's surface is wrapped in the awesome presence of the sea, and the ocean guards its secrets well. Like a half-silvered mirror, it reveals glimpses of the world below, but then it hides them in ripples like laughter. So, it is not surprising that everyone loves mysteries and that they live on despite any evidence that seems to disprove them.

One of the longest-lived mysteries is the story of the lost island of Atlantis. The story was introduced to the Western world by one of our most honored philosophers, a fact that has contributed to its long life. Plato described a legendary island in the Atlantic Ocean west of the Columns of Heracles, or Pillars of Hercules, two promontories in the Straits of Gibraltar, named for the Greek hero Heracles (Latin form, Hercules). The island was swallowed up by the sea in an earthquake and deluge that lasted one day and one night. This story had been told to the Athenian poet and lawgiver Solon by Egyptian priests in the sixth century B.C., and it was repeated by Plato in the fourth century B.C. in two of his dialogues, *Timaeus* and *Critias*. It was the site of a rich and powerful kingdom with a level of civilization far exceeding that of other lands during this period of history—9000 years before the time of Solon:

> There was an island situated in front of the straits which you call the Columns of Heracles: the island was larger than Libya and Asia put together, and was the way to other islands, and from the islands you might pass through the whole of the opposite

continent which surrounded the true ocean; for this sea which is within the Straits of Heracles is only a harbor, having a narrow entrance, but that other is a real sea, and the surrounding land may be most truly called a continent. Now, in the island of Atlantis there was a great and wonderful empire, which had rule over the whole island and several others, as well as over parts of the continent; and besides these, they subjected the parts of Libya within the Columns of Heracles as far as Egypt, and of Europe as far as Tyrrhenia. The vast power thus gathered into one, endeavored to subdue at one blow our country and yours [Egypt and Greece], and the whole of the land that was within the straits; and then, Solon, your country shone forth. . . . She defeated and triumphed over the invaders, and preserved from slavery those who were not yet subjected, and freely liberated all the others who dwelt within the limits of Heracles. But afterward there occurred violent earthquakes and floods, and in a single day and night of rain all your warlike men in a body sunk into the earth, and the island of Atlantis in like manner disappeared, and was sunk beneath the sea. . . . Many great deluges have taken place during the nine thousand years, for that is the number of years which have elapsed since the time of which I am speaking.

In *Critias*, Plato describes in detail how "this sacred island lying beneath the sun brought forth fair and won-drous [things] in infinite abundance . . . flowers or fruits grew and thrived in that land." Food was abundant and many animals including elephants lived there. Bulls were raised and used for religious sacrifice. There was wood for the carpenters and metals lay buried in the earth. In this great and powerful land there were tem-ples, palaces, harbors, docks, and canals. The public

buildings were ornamented with statues of gold and pinnacles covered with silver. The island was divided into
10 kingdoms, each ruled by a king who had absolute
control over the citizens, punishing and slaying at will.
Within each kingdom, there were local leaders who were
required to furnish their king with armed men and provisions for war. There was a written language; decisions
and laws were inscribed on golden tablets and on the
walls of the temples.

From Plato's description, Atlantis enjoyed a very
favorable environment—a warm climate where elephants could thrive. Abundant crops provided food for
a large population. The human society was very
advanced, and the written word had been invented. The
people were well organized and skilled in warfare, having
conquered most of the known world.

It is no wonder that those who read about Atlantis
wanted to claim descent from this successful society, and
everyone was anxious to discover the remains of this
wealthy civilization beneath the sea.

In Plato's time the Greeks had not explored beyond
the Strait of Gibraltar, so this part of the Earth was
unknown to them, and they had no way of judging the
likelihood of a very large island having existed in the
Atlantic, nor did they have an understanding of the way
climate and even the shape of continental masses and
ocean basins change over time.

In the centuries that followed Plato, many theories
were advanced to identify the drowned island of
Atlantis. Medieval writers, hearing the tale from
Arabian geographers, believed it was true, and they also
believed the accounts of other islands in the western
sea: the Greek Isles of the Blest, where the spirits of
heroes went after death; the Welsh legend of Avalon;
the Portuguese Antilla or Isle of Seven Cities, and St.
Brendan's Island, which was the subject of sagas in
many languages. All of these except Avalon were

marked on maps of the fourteenth and fifteenth centuries and were the object of voyages of discovery. Somewhat similar legends are those of the island of Phaeacians, mentioned by Homer; the Island of Brazil, a sunken island off the Cornish coast; the lost Breton city of Is; the French Isle Verte; and the Portuguese Green Island. The last appears in many folk tales, from Gibraltar to the Hebrides, and until 1853 it still appeared on English charts as a rock at 44°48' north latitude and 26°10' west longitude.

Serious attempts were made to identify evidence of Atlantis. Some thought the Basques, whose ancestry has never been discovered, were descended from Atlanteans who escaped by boat when the island was inundated and landed on the northwestern shore of Spain. (There the Basques have maintained their own distinctive culture with a language and customs unlike those of any other nation.) Or perhaps the Guanches, members of an ancient tribe that originally settled the Canary Islands, were survivors of the catastrophe that destroyed Atlantis. Their descent has also been a mystery. In these latter cases, the demise of Atlantis could resolve other historical puzzles, because these surviving people occupy the general area where Plato had located Atlantis. Other, more widely scattered lands—the West Indies, the Azores, Sweden, and Ceylon, have also been proposed as the remnants of the great land that had been swallowed by the sea.

In the late nineteenth century, during the laying of the transatlantic telegraph cable, the existence of a large ridge was discovered in the middle of the Atlantic Ocean. Named the Dolphin Ridge, its discovery was used to support the theory that a continent had existed there, and its remains now lie on the bottom of the sea. This theory was expounded in *Atlantis: The Antediluvian World* by Ignatius Donnelly (1884). Atlantis, Donnelly stated, was a large continent situated in the center of the Atlantic Ocean, and it was inhabited by a superior race until about 11,500 years ago, when it was swallowed up by the sea. The survivors made their way to other lands, taking their superior culture with them, and this knowledge was the basis of the important early civilizations of mankind—the Egyptian and Greek, as well as the cultures of the new world, the Mayan and the Incan. Donnelly's book was very popular and became the core of a persistent cult. Unfortunately, the book was flawed with many inaccuracies and misinterpretations of geological and human history.

Scientific advances have produced facts that cast serious doubt on Donnelly's theory, as well as many of the other explanations of the myth of Atlantis. The underwater world has been explored by wonderful new techniques, revealing an incredibly diverse land lying below the surface of the sea. Specially equipped research vessels have crisscrossed the oceans, taking cores from the sea bottom. These cores contain valuable information about the crust of the Earth beneath the sea: the mode of its formation; its age; the temperature of the surrounding waters; and even the magnetic field when the rocks were formed. Where the depths are too great to be cored, instruments towed behind research vessels have recorded reversals in the magnetic field.

Surprising as it may be, the direction of the Earth's magnetic field has reversed itself throughout the history of the planet every million years or so. When molten magma solidifies in the cold ocean waters, the rock is magnetized in the direction of the field at that moment in time. These reversals provide an important record of the rate of seafloor spreading, the creation of ocean basins, and the movements of pieces of the Earth's crust around the world.

Portions of the ocean floor have been directly observed by men in such submersible craft as the *Alvin*. Photographs and samples taken during these dives have provided more detailed information, such as the existence of deep-sea vents, like those discovered near the Galápagos Islands.

When all these data are put together, the picture of an amazing landscape emerges. Far from the simple image of a uniform plain surface, reflecting the relatively flat surface of the ocean's face, the deep-sea bottom is a scene of soaring mountains and deep canyons, a grandiose landscape of craggy black cliffs and gently rolling hills dusted with "snow." This white powder is a fine layer of sediment composed of the tiny shells of trillions of microscopic organisms that drift down onto the seafloor day after day, year after year, and collect on the high slopes. At greater depths, where the water is colder and more acidic, they dissolve in the seawater. The accumulation occurs very slowly—less than a tenth of an inch in a thousand years.

The most impressive feature of this hidden land is a tremendous mountain range that bisects the Atlantic Ocean and continues in a broad loop between Antarctica and the Cape of Good Hope, up through the center of the Indian Ocean, then down around Australia and through the South Pacific to Mexico. It is almost 30,000 miles long and it averages 600 miles wide, occupying nearly one-fourth of the entire surface of the planet. Iceland lies at the northern end of this rift. The area called the Dolphin Ridge is a very small segment of the Mid-Atlantic Ridge and the rocks on its crest are very young.

At the top of the mountain range there is no "snow"; dark black rocks stand out prominently along the highest peaks. At close range it is apparent that these shining black masses are composed of basalt, one of the rocks that characteristically form from molten lava when it cools rapidly. Rocks like these have been observed to form in places such as Hawaii, where lava erupts beneath the sea and solidifies quickly in cold water, preserving their extruded shape like ribbons of toothpaste squeezed from an enormous tube.

When samples taken from various locations on the rift were dated, it was discovered that the rocks at the top are among the youngest pieces of Earth's crust. They have not existed long enough even to collect a layer of sediment. Those from the center of the rift are usually less than 1000 years old, and the rocks become progressively more ancient in places farther from the ridge. New crust is actually being formed along the entire crest of the enormous underwater mountain range.

Deep-sea drilling has revealed the surprising fact that the ocean floor is almost entirely volcanic in origin. The underlying base below any accumulations of sediment is composed of basalt. Furthermore, the ocean crust is relatively thin. It is only 4 to 5 miles thick, while the crust of the continents typically varies from 20 to 35 miles in thickness. The ocean floor is being continuously created in the great rift system and swept back again into the soft interior of the planet in trenches like those along the ring of fire around the Pacific. This action takes place rapidly enough to have made anew the entire ocean floor—about two-thirds of the planet's surface—twenty times since the Earth was born. Rocks 3.8 billion years old have been found on land—in Greenland, for example— but no sample of ocean floor older than 180 million years has been discovered. In geologic perspective, 180 million years ago is the comparatively recent past. At that time mammals had already evolved and had taken up residence on every continent. This was the heyday of the enormous reptiles. *Brontosaurus, Stegosaurus,* and *Brachiosaurus* roamed the land masses of the Earth. So, the crust of the ocean floor is younger than the bones of many dinosaurs that we see displayed in museums around the world.

Between the edge of every continent and the true ocean floor there is a shelf of land rising about a mile above the sea bottom. This is the real edge of the continent. The shelf is relatively flat on top—gently rolling land with low hills that resemble sand dunes on the desert. As sea levels have risen and fallen throughout geologic time, traces of old beaches—deposits of gravel and sand—have been left on the shelves. Fossilized bones of ancient land mammals—mastodons, mammoths, and giant sloths—have been uncovered there, as well as beds of peat containing fossils of grasses, twigs, and pollen. This testimony from the past confirms the assumption that the continental shelves are really drowned portions of the continents.

Cores taken through the sediment and rock show that the shelves lie on a granitic base like the continents, not on the basaltic crust that is characteristic of the ocean floor. Over these rocks lie great accumulations of soil washed down from higher land.

All the continents on Earth are surrounded by shelves that vary greatly in width and depth. The depth may be anywhere from 30 to 2000 feet. The width varies from less than half a mile off the west coast of South America to more than 800 miles off Siberia. All together, the continental shelves occupy one-tenth of the world's surface.

Taking all these facts into consideration, it seems apparent that the existence of a continental land mass that broke up and sank to the bottom of the Atlantic Ocean would have been discovered. In fact, even a drowned continent the size of Asia Minor and Libya in any of the oceans would almost certainly have been detected, although something much smaller like an island might still be hidden beneath the surface of the sea.

There are other facts, however, that are difficult to reconcile with the description presented by Plato. Timing is the most outstanding problem. Plato said in *Timaeus* that Atlantis was swallowed up by the sea 9000 years before the time when Solon was talking with the Egyptian priests. The climax of Solon's career came in 549 B.C., when his laws were adopted by the Athenians, approximately 2600 years ago. Adding 9000 years to that date puts the demise of Atlantis at 11,600 years before the present time. That was a very interesting moment in geologic history. It was near the end of the last great ice age. For approximately 3 million years, glaciers had advanced and retreated across the land, and as this happened, sea levels rose and fell. Most of the changes, however, occurred very slowly—over many centuries— and may not have been perceptible to the human eye. But toward the end of the last glacial period, there were several short-term oscillations—the most significant between 11,000 and 12,000 years ago. Those changes may have occurred rapidly enough to have been significant in a human lifetime.

Archaeologists have found evidence of the suddenness of this climatic change. A buried forest has been discovered in northeastern Wisconsin on the shore of Lake Michigan. There, dozens of large spruce trees had been toppled and drowned. They lie buried in lake sediment and are covered by later deposits typical of another glacial period. Radioactive dating indicates that these stumps and logs are 11,850 years old. This finding is so significant that the name of this place, Two Creeks, Wisconsin, is also applied to this warm interval. The period corresponds roughly with a similar episode in Europe known as the Allerod glacial retreat, which took place 11,000 to 12,000 ago.

Submarine topography and dredging indicate that important changes in sea level occurred in the North Sea during this same period. A plateau nearly as large as Denmark now lies about 60 feet underwater, sloping abruptly at its edges into much deeper water. This topography was discovered by fishermen, and the raised area was named the Dogger Bank. The fishing trawls brought up a strange assortment of the remains: fragments of trees, bones of large mammals, and crude stone instruments.

When these finds were studied by scientists, an interesting history of the Dogger Bank emerged. During the ice age the entire floor of the North Sea was emergent, although much of it may have been a low wetland covered with peat bogs where mosses and ferns grew. Gradually, forests from the neighboring higher lands moved in—willow trees and birches. Animals came from the mainland. There were bears, wolves, hyenas, wild oxen, bison, woolly rhinoceros, and the mammoth. Humans followed and hunted the animals, using arrowheads and axes chipped from flint.

Then, for reasons that are not clearly understood, the climate became warmer. The glaciers began to melt and floods poured into the sea. For a while the highest part of the continental shelf remained above sea level. Dogger Bank became an island. But, as the melting continued, the island was covered by the sea. We can imagine that the men who escaped and made their way to the mainland told a frightening tale of the great flood that had swallowed up their land. In the manner of human nature, they may have exaggerated the comforts and opulence of their cities on the lost island. These stories would have been passed from generation to generation. With each telling the tale grew, until many centuries later, when it was retold to Solon by the Egyptian priests, who may have added details gleaned from the stories of more recent disasters and lost civilizations. Thus were the cities of Atlantis embellished with palaces and statues of gold, and in this way the legend of Atlantis may have been created.

But the advanced civilization depicted by Plato (and as elaborated by Donnelly) does not seem credible as long ago as 11,500 years. The history of human development indicates that during that period of time mankind was still in the Stone Age. Mining and metalworking were accomplishments that came much later. The Bronze Age cultures did not begin until about 5000 or 3000 B.C.

If indeed Atlantis was a settlement built on a continental shelf more than twelve millennia ago, we must assume that it was a Stone Age culture—although possibly an unusually successful one. By this time a few human communities were making the transition from savagery to a more settled way of life based on animal husbandry and agriculture. Population had increased, and this change led to the beginnings of town life and social and political organization. Fire had been discovered, as well as the art of using clay for pottery. Woven fabrics began to take the place of animal skins, and sometime during this transition period the first boats were fashioned. We also know that human beings at that time and even much earlier were capable of great skill and artistic sensitivity, as demonstrated by the paintings at Lascaux cave in France and Altamira cave in Spain (13,000 to 15,000 years ago).

However, many of the details described by Plato could not possibly portray a Stone Age community: city walls covered with a coating of brass, palaces, and temples surrounded by enclosures of gold, roofs of ivory adorned with gold and silver, chariots borne by six winged horses, and inscriptions on the columns of the temples (the earliest examples of the written word date from the fourth millennium B.C.). These and many other details of the Atlantis story describe a late Bronze Age culture of great wealth and power.

One attempt to explain this discrepancy was proposed by A. G. Galanopoulos. Beginning in 1960 he published a series of articles supporting the hypothesis that the eruption of Santorin was the event that led to the myth of Atlantis. He had an interesting and novel suggestion. Perhaps, he said, an error was introduced when the tale was communicated to Solon by the Egyptian priests. One of them mistranslated the Egyptian word or symbol for 100 as 1000. This error would have increased by a factor of 10 all the figures over 100 in Plato's account, whether they referred to time, dimensions of area, or numbers of men and ships. A correction of this mistake would reduce the figures to values that are consistent with what is known of the Minoan civilization. For example, 9000 years becomes 900, and when that number is added to the age of Solon when he traveled in Egypt, the date falls in the middle of the second millennium B.C., approximately the time of the eruption of Santorin. In many ways, the description of Atlantis given by Plato is similar to the Minoan culture—a seafaring nation, stronger than any of its neighbors, where bulls were raised and used for sacrifice. There was a sumptuous royal palace (like Knossos) of great size and beauty. Art and architecture were highly developed. There was a written language.

However, there are some details that do not fit the Minoan history. Santorin does not lie beyond the Pillars of Hercules. Although the influence of the Minoans did extend throughout much of the Mediterranean and they apparently exacted tribute from the Greeks, they were not a warlike people and they did not conquer most of the known world. Their cities were not surrounded with walls clad in brass. Finally, the natural disaster did not totally wipe out Santorin. Five much smaller islands remained.

All of these differences in the story can be explained as inaccuracies resulting from the fact that the eruption of Santorin and Plato's description of Atlantis were separated by about 1000 years. There is no written record of the event. When the Minoan civilization was in full flower, the Greeks did not yet have a written language, and the early Egyptian hieroglyphics did not reveal any facts about the Minoans. So, the impressions were passed down by word of mouth, and now it is impossible to tell how much of the story is due to the imagination of the generations of storytellers who reworked the tale.

The same argument (of exaggeration and embellishment) can also be used to support the theory that Atlantis was the site of a Stone Age culture that was drowned in a flood 11,500 years ago. The difference is just a matter of degree.

But the identification of Atlantis with Santorin is considerably less interesting than Plato's tale. It adds no new dimension to history; the details of the Minoan culture are already quite well known. Therefore, to many people this is not a satisfactory denouement of the mystery.

—

Mankind does not want to give up fairy tales. They inspire poems and sagas and voyages of adventure. Even with the knowledge we have gained about the ocean floor and the history of the planet, there is still room for discovery. The evidence of an ancient island may even now lie hidden beneath the enigmatic surface of the sea.

The fact that this can happen is demonstrated by the discovery in 1997 of the remains of the ancient royal quarter of the fabled city of Alexandria, which contained Cleopatra's palace. Sixteen hundred years ago this quarter slipped into the sea, where it has been ever since, just 300 feet from the present shoreline in water only 15 feet deep. This harbor had been searched before, and nothing was found because the water was nearly opaque with pollution. The ruins were buried deep beneath sand and encrusted with thick layers of residue from untreated sewage that has been emptied into this harbor from the city that now contains 5 million people. French archaeologists and divers using modern technology have been able to penetrate the gloom and identify statues, gigantic granite blocks, and columns bearing Greek inscriptions.

As the scientists have persisted in their search, many treasures have been uncovered in their watery graves, some from a much earlier time. Twenty-six sphinxes were found, several dating back to the sixth century B.C., as were pieces of an obelisk with Egyptian inscriptions. Alexandria was founded by Alexander the Great in 332 B.C., and at that time Greek was the official language. Therefore, the stones with Egyptian hieroglyphics and the most ancient sphinxes were objects left from a more ancient time. This fact suggests that an earlier city stood on the same site.

The little island of Pharos in the harbor contained a famous landmark—the lighthouse of Alexandria, which was known as one of the seven wonders of the ancient world. Estimates of the height of this stone tower range from 200 to 600 feet. It was built in three stages, all sloping slightly inward—the lowest was square, the next octagonal, and the top cylindrical. A broad ramp led to the top, where a fire burned at night. A mirror reflected this light, making a bright beacon visible far out at sea, leading ships into this most important city in the ancient world.

Enormous granite blocks, so massive that they could never have been moved as single pieces, lie on the floor of the harbor, and when these are reassembled in the imagination, they make a structure of such size and importance that it might well have been a wonder of the ancient world. Pieces of two giant statues were also found nearby. If they could be reconstructed they would stand 40 feet tall—male and female (king and queen)—and may have stood at the entrance to the harbor.

Ancient accounts have described the royal quarter of Alexandria as a walled city with splendid, lavish buildings and beautiful parks, the whole making up about a third of the total area of the city. The period of magnificence for Alexandria ended in 30 B.C., when Cleopatra's army was defeated in the Battle of Actium and Egypt became a Roman province. In the fourth century A.D., a great earthquake and tsunami struck the city, and the royal quarter vanished beneath the sea. But the lighthouse—known as Pharos of Alexandria—was still standing in the twelfth century and was finally destroyed (probably by another earthquake) in the fourteenth century.

These recently discovered underwater treasures add credibility to the story of a much more ancient sunken city and keeps alive the tale of Atlantis.

The legend of a lost civilization has had an enduring hold on the imagination of mankind. It is not just the tragedy of great wealth and glory swept away in a single day and night. It is something much more fundamental— the dream of restoring a belief in ourselves.

The presence of a mysterious culture that was said to have been very advanced awakens the hope of discovering unrealized potential in human beings. Perhaps the inhabitants of this civilization took a path different from ours—a path that did not lead to weapons of mass destruction and victory in war. Perhaps they had struck gold, not buried in their hills but in human nature—creative gifts more significant than any realized in our culture. Exactly what these gifts are we cannot know, but it is important that we keep our belief in ourselves and our sense of wonder, because the reality of the universe and life have dimensions greater than any we have ever dreamed of.

Chapter 14

THE EARTH

An Island in
the Universe

The greatest beauty is organic wholeness,
The wholeness of life and things.

—ROBINSON JEFFERS

The Space Age has given us a great gift—the ability to see the Earth as a whole, a beautiful blue bubble set in the black oceans of space. It is indeed an island in the universe. It is also isolated, because approximately 240,000 miles separates it from the nearest significant assembly of matter (the moon) and 93 million miles separates it from the nearest star (our sun). This little island harbors at least 10 million forms of life that it feeds and nurtures. It is a reasonable assumption that all the life forms we see on Earth are endemic. They have evolved here, and they probably do not exist in exactly this same form anywhere else in the universe. The planet possesses great beauty clothed with mystery and wonder—the shining mountain ranges iced with snow, the everchanging blues of sea and sky, and the green and golden grasses rippling in the wind. It holds a multitude of exquisite little things, such as the unfolding shape of an opening flower or the dance of fireflies on a summer night.

As with individual islands, the planet has a past, which we are beginning to understand, even from its birth four and a half billion years ago. At that time, the infant Earth emerged from the whirling cloud of the solar nebula—a diffuse ball of gases and dust enveloped in a kind of afterbirth, the primordial atmosphere of very light gases. Gradually, this envelope drifted away into space, and, at the same time, the material of the embryonic planet was drawn strongly toward the center under the force of gravity. So much heat was generated by this "falling of matter" that the planet became very hot, and the elements and the mineral compounds melted. The densest ones flowed to the center and the less dense ones rose to the top, like foam in a boiling kettle. The surface of the compacted ball was constantly bombarded with more matter from space as remaining particles from the solar nebula were attracted by the gravitational field. But still the planet cooled; the solid crust began to take shape.

Even as the planet formed, it was disrupted by violent eruptions and incandescent flows of glowing lava. Gases were emitted; clouds of water droplets condensed in the atmosphere, and rain fell; oceans began to collect on the planet's surface, and soon living molecules were present in its shadowy depths. Exactly when they first appeared we cannot say. It is a phenomenon of great importance, however, because the creation of these earliest forms of life involved a higher degree of organized activity than anything that had previously occurred on our planet. The simplest self-replicating organism contains hundreds of thousands of atoms engaged in multitudinous coordinated activities. And all of this occurred in the first billion years, or aeon, after the birth of the planet.

By the early years of the second aeon, the first cellular organisms appeared, and photosynthetic algae were forming vast colonies in the sea, altering further the composition of the fluid membrane that wrapped the planet. Throughout this whole process, the body of the Earth was never still. It constantly readjusted and reorganized the materials of which it is composed. The very skin of the planet was continuously reworked and made anew, like a snake's skin grown and cast off many times to make room for a new one. The ocean floor has been melted down and reformed about five times every aeon. The dry land has been transformed, too, as mountains were lifted up, then swept away by wind and rain. And while all of these dramatic movements were occurring, less conspicuous changes were quietly proceeding beneath the surface. Deposits of copper and gold were secreted in the deep roots of volcanic mountains, and beautiful crystals took shape inside the rocks—diamonds and star sapphires and blood-red rubies, like those that lie scattered in the soil of Sri Lanka.

As in the creation of an atoll, the early years of island Earth involved volcanism and the formation of a symbiotic relationship with living things. The planet was continuously altered by the presence of life. By-products of living forms in the sea changed the composition of the atmosphere, and they in turn were affected by the flow of gases and heat from hydrothermal vents in the ocean bottom, which poured forth their hot, mineral-laden waters. Life soon began to invade the land and profoundly altered the surfaces. Throughout the next two and a half aeons, floods of living things appeared and disappeared, but life itself continued to burgeon, moving to ever higher, more organized beings. And these beings were developing symbiotic communities that influenced the response of the planet to all the forces that flow in upon it. Much as a coral reef, the Earth can be viewed as a superorganism—at the very least, one in the process of formation—a thought eloquently expressed by Lewis Thomas in *The Lives of a Cell:*

> Viewed from the distance of the moon, the aston-
> ishing thing about the earth, catching the breath,
> is that it is alive. The photographs show the dry,
> pounded surface of the moon in the foreground,
> dead as an old bone. Aloft, floating free beneath
> the moist, gleaming membrane of bright blue sky,
> is the rising earth, the only exuberant thing in this
> part of the cosmos. If you could look long enough,
> you would see the swirling of the great drifts of
> white cloud, covering and uncovering the half-
> hidden masses of land. If you had been looking for
> a very long geological time, you could have seen the
> continents themselves in motion, drifting apart on
> their crustal plates, held afloat by the fire beneath.
> It has the organized, self-contained look of a live
> creature, full of information, marvelously skilled
> in handling the sun.

The British scientist James Lovelock turned this vital interpretation of the Earth into a well-supported theory.

He believes that the entire biosphere, together with the atmosphere, the oceans, and the soil, is a self-regulating system maintaining the conditions in which life can flourish. He named this entity "Gaia," after the Greek goddess of the Earth. Evidence for the existence of this system can be found in the extraordinary chemical composition of the atmosphere and the oceans. The proportion of the gases in the air is "improbable by at least 100 orders of magnitude," Lovelock says. And yet this same combination of elements—extremely favorable for life—has been maintained within very narrow limits over the aeons when photosynthetic vegetation was present on land.

The same argument, Lovelock maintains, can be applied to the composition of the oceans. A salt concentration of more than 6 percent would be lethal for almost all marine organisms. The average saline content today is 3.4 percent, and there is geological evidence that this figure has not varied significantly for three and a half aeons, although all the rivers have been continually emptying more salt into the sea. These facts suggest that the salt level is under some kind of central cybernetic control.

The uniformity of the Earth's climate is another fact that demands explanation. With the exception of a few relatively brief ice ages, the temperature range has not changed appreciably since life first appeared on Earth. Water in the liquid state existed on the planet at that time, even though astronomical theory tells us that in those early years the energy radiated by the sun was at least 30 percent less than it is today. Thirty percent less heat from the sun would imply a mean temperature for the Earth well below the freezing point of water. If the Earth's climate was determined solely by the output from the sun, our planet would have been in a frozen state during the first one and a half aeons of life's existence. "We know from the record of the rocks," said Lovelock, "and from the persistence of life itself that no such adverse conditions existed. ... It was warm and comfortable for embryo life, in spite of the weaker flux of heat from the sun."

Various phenomena have been invoked to explain how all these favorable conditions could have occurred by chance and been sustained by a series of coincidences. But as a series of lucky circumstances grows longer, it becomes more and more improbable. The same facts can be explained more reasonably, Lovelock believes, as the consequence of a self-regulating system that is actively manipulating the environment and maintaining the conditions most favorable for its own existence. Even after great extinctions damaged large portions of this delicate web of life, it regenerated and flowered anew. The living Earth system has taken an active part in maintaining and extending its own existence. It is not just passively pushed around by external forces. In all these characteristics we recognize an integrated whole—perhaps even, as Lovelock suggests, a living organism. Life is a stage in the organization of matter; the point where an organism becomes a living thing is simply a matter of definition.

The appearance of *Homo sapiens* represented a quantum jump toward higher levels of organization, because even the earliest human mind was the most elaborate and delicately balanced integration of matter known in the universe. And—at first—the ancestors of man lived in close collaboration with the rest of nature. Trees provided their first home. High off the ground, they offered protection from predators in their leafy branches. By climbing and leaping from tree to tree, primates acquired agility, strong arm and leg muscles, and stereoscopic vision. When the weather changed and forests gave way to open savannas, these primitive men were able to walk upright, freeing the hands to pick up and manipulate objects and to fashion tools.

In the meantime, new and much more effective methods of reproduction were beginning to occur in the green vegetation that clothed the Earth. Seeds, which were an improvement over the naked spores, were vulnerable to the vagaries of wind and ocean currents and the nature of the land on which they fell. But about a 100 million years ago, nature invented a more advantageous method of packaging the seeds, improving their chances of survival, as we noted on our visit to the Hawaiian Archipelago. In these angiosperms, the seeds are wrapped in an additional layer of covering. Some of these coats are hard, like the shell of a nut, providing extra protection. They can survive, lying dormant for long periods of time. Some can float on the ocean currents, reaching new habitats where the conditions may be most favorable for growth. In other angiosperms the coats are soft and sweet-tasting; fruits like bananas and cherries entice birds and other animals to eat the flesh and carry the seeds to different environments. These new potentialities conferred biological advantages, and angiosperms thus spread rapidly throughout the world. They also speeded the evolution of many other forms of living things.

The angiosperms provide food so rich in energy that most animals need to spend only a portion of their lives

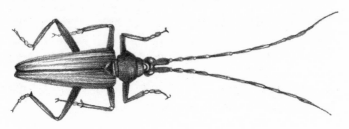

in obtaining nourishment. They no longer have to depend on thin sustenance, as does the koala, who still lives in trees and eats eucalyptus leaves all day. For mankind, the time thus released made possible some of the great human achievements, such as the invention of writing and the beginnings of art and music.

When a primitive man first picked up a handful of seed and sensed its potential, as Loren Eiseley said, "In that moment, the golden towers, his swarming millions, his turning wheels, the vast learning of his packed libraries, would glimmer in the ancestor of wheat, a few seeds held in a muddy hand. . . . The weight of a petal has changed the face of the world and made it ours." Most of this evolution proceeded in a positive direction, opening up new opportunities and discovering new talents that had been hidden beneath the primal demands of existence. These changes have been achieved by the close cooperation of all the myriad aspects of nature. But as time has gone by, mankind has started to disturb this balance, and examples of the results are demonstrated in the history of island communities.

Islands provide very useful information about the interaction of various forms of living things. Islands are small, encapsulated units that can be studied and understood in a way that is not possible when we are dealing with the complex interwoven relationships of life on a large continental land mass.

One of the most interesting facts we have learned about islands is that they have given birth to many fascinating endemic species, including the coral trees (wili-wili) and white hibiscus that grow wild in Hawaii; the Indri lemur, whose song echoes through the forests of Madagascar; the clumsy dodo bird, which lumbered across the hills of Mauritius; and the mysterious coco-de-mer that survives in the Valley de Mai in the Seychelles. The list goes on almost without end.

The explanation for these island treasures is not hard to find. The conditions were especially favorable on islands for new forms of life. Everywhere nature is prodigal in inventing new species and subspecies. Most of these have little chance of survival in continental environments, but on very young islands there are fewer predators, much less severe competition, and many still unoccupied niches.

There are small environmental differences even among islands in the same archipelago. In the Galápagos, for example, the ground finches evolved to take advantage of the opportunities offered by each island habitat. The ancestral ground finch was probably blown in by the wind from the mainland of the Americas and landed in an area almost free of competition for food or territory. There were no predator enemies like hawks or eagles. This isolation facilitated the evolution of new species, unique to each habitat. Throughout many generations the different groups of finches formed distinctive lifestyles. Some developed beaks that were efficient for cracking nuts, others for catching insects or for feeding on fruits and flowers. One finch even learned to use a cactus spine to probe grubs out of holes in tree branches. Thus, an original single stock had deployed into an array of separate and different species. Charles Darwin observed these innumerable trifling details of structure, and this observation was the spark that ignited his important theory—that life has evolved by natural selection, the survival of the fittest.

Even when islands lie quite close together—as they do in the Galápagos—there seems to be little tendency to take over the neighboring habitats. There is no competition, for example, of the finches on Albermarle with those on James Island. Given sufficient food and territory, the related species do not compete for each other's home base. Evolution has carried them far enough apart that the invading species would be at a competitive disadvantage in an unfamiliar habitat. As a result, the various species remain separated and evolve in their own special ways.

We saw this same phenomenon on Bali and Lombok. Across Wallace's Line, there was no intermingling between the Australian and the Asian fauna, even though these two islands are in sight of each other across this invisible line.

In some cases, large continental land masses have acted like islands, providing isolated conditions under which animals and plants are protected from competition with other more efficient species. About 90 million years ago, the splitting up of Gondwanaland left Australia an isolated land mass, cut off by the sea from the rest of the world. By this time primitive marsupial mammals had colonized Australia. Marsupials give birth to very immature offspring and usually carry them in pouches. Placental mammals, which nurture their young to a more advanced stage within the mother's body, had not yet appeared there. So, the marsupials had a free field and they radiated into many species, some of which are found nowhere else in the world, for example, leaping kangaroos, phalangers, bandicoots, and the charming little koala with round eyes and fluffy ears.

Placental mammals, however, are more effective animals than the marsupials, because they have larger brain capacities and better temperature regulation, among other biological advantages. When placentals like the rabbit and the dingo were introduced into Australia, they multiplied explosively. Dogs, cats, goats, and cattle were brought by colonists. Rats came ashore from the ships, and eventually many of the marsupials became endangered. The most disruptive invader was man himself—man the predator, endowed with a mind that gave him vast advantages over simpler forms of life. He could make traps and use arrows and slings and later guns to harvest the native animals for food—or sometimes just for sport. The clumsy kangaroos were hunted almost to extinction; the koalas were captured for their fur and many died because of human encroachment on their eucalyptus forests.

In this way, on islands around the world, which, by their isolation, had nurtured relatively vulnerable native species, the communities were disrupted by the introduction of alien and more successful species. Therefore, the incidence of endangered or extinct species is greater on islands than on continents. More endemic species have been created on islands but more have perished there.

As better and swifter modes of transportation have been invented, there are no islands anyplace in the world that are truly isolated anymore. Even the remote islands in Indonesia and the South Pacific are regularly visited by cruise ships and aircraft. Cruise lines frequently advertise "spend a day on an uninhabited island," and then land boatloads of 1000 tourists on a little spit of land. Even Easter Island has several flights that bring visitors every day. Naturalists have good reason to be concerned that diversity is diminishing and that there are fewer places where new species are being created. But it is unrealistic to suggest that human beings will curtail their travel. We must look for other ways to reduce our impact on nature.

The presence of isolated islands has endowed us with a wealth of endemic species that add variety to the biotope. This is an important contribution, because whenever the environment changes, species must adapt to the new conditions or become extinct. The existence of many small variations between species and subspecies increases the probability that at least one of these will make a satisfactory adaptation.

But in our tightly integrated "one world," isolated communities are rapidly disappearing. Consequently, the creation of new endemic species is also reduced. To preserve all the endemic species that now exist will not be possible, because the invasion of mankind and technology inevitably changes the environment in which these species evolved. Careful attention to reducing our impact can help but not completely eliminate the reduction. And all of this is happening very rapidly.

Given the situation, it is imperative that we study the endemic species that are present today in each island community and identify those that have special properties, which can contribute in some important way to our lives—for example, the pink primula in Madagascar and the baobab tree whose wood resists invasion by termites. There are undoubtedly many other species whose potential has never been recognized. Unless we act quickly, these will be gone forever. We might think of them as the buried treasures of gold and gemstones that nature has stored up in certain favored places around the world. But those gifts are not endangered by the passage of time, while many endemic species are traveling down the road to extinction.

In some cases ecosystems that harbor endemic species have been protected by the creation and preservation of sanctuaries. In the Galápagos and several of the Seychelle Islands, birds, land tortoises, and unusual plants are sheltered from careless interference. But these are only a few isolated examples. How complete and how dedicated are these attempts at protection? Remember Johnston Atoll, where the bird preserve now shares a tiny motu with incinerators for chemical weapons.

Many environmental projects today concentrate money and resources on attempts to save particular species from extinction. I am reminded of the pink pigeon and the kestrel in Mauritius. Although these attempts have been initially successful, they may not be in the long run, because they do not address the underlying causes—the destruction of the closely integrated community of living things that was the natural environment of these species. The forests with their particular varieties of trees affected the climate, the water supply, and the soil. The presence of small wild fauna on the forest floor provided food for the kestrel. There were berries and other fruits for the pink pigeon. Even the careful provision of these supplies cannot duplicate the experience in nature. And the use of artificial insemination raises concerns about the dangers of inbreeding.

There are natural minimum limits to the size of a successful species. As we have seen on Pingelap, genetic inheritance can be weakened by inbreeding. A recessive and undesirable gene may be reinforced by combination with a similar gene from the other parent, and this combination may be expressed as a disability that undermines the health and well-being of the individual. As breeding populations are reduced in size, the chances of duplicating undesirable genes increase and the vitality of the species is reduced. This is one of the factors that drives a species (animal as well as human) toward extinction.

A factor that is much more difficult to quantify is the spiritual strength that comes from being part of a successful and smoothly functioning community. In our review of island life, we have seen the great differences that result from the character of the communities that have evolved in relative isolation. In Mauritius and Lombok, for example, the cooperation between many different segments of the societies, with remarkable racial and religious tolerance, has resulted in economic success and a peaceful and pleasant lifestyle.

In the Bahamas, Harbour Island endured a direct hit by the worst hurricane of the century and the people came out stronger and more self-assured than they had been before. Black and white, rich and poor, they all cooperated and helped each other to meet this challenge. Here again was a society that worked.

But there are other, less successful stories. The dreadful history of Easter Island is an outstanding example. On this tiny island, severely separated from other land masses, two disparate factions, the Hanau Eepe and the Hanau Momoko, shared a habitat but never forged a true community, never learned to treat each other with tolerance and respect.

Problems arose when the human population grew beyond the size that could be comfortably sustained in this limited environment. Most species have natural controls that help maintain a favorable balance between the number of individuals and the resources of the environment. Parrot and butterfly fishes trim excess growth from the coral reefs, and sharks and barracuda keep the fish population within reasonable limits. Several types of rodents, like squirrels, muskrats, and lemmings, reduce overpopulation by migration. The juveniles move away from a home area that has become too crowded and where food has become too scarce.

Human beings also migrate under

these conditions. The traditional pattern of land use has been cut, burn, plant, destroy, and move on. However, such a system is only possible where people have plenty of room to move from one place to another. In isolated places like Easter Island, this is not a viable solution. People on Easter Island sought different ways to overcome the natural controls and extend the resources. The use of fire offered a solution to the problem. When energy and food became too scarce to support a growing population, trees were cut down to clear land for raising more crops. The wood provided fires to warm the dwelling places and cook the food. Thus released from the normal restraints, the human population on Easter Island continued to grow, and eventually most of the forest cover was stripped off the island. Then the land surfaces dried out more rapidly and the precious nutrients were leached away. The land, thus desiccated, was not able to produce large crops. But the number of people continued to increase. Eventually, hunger drove the two warring factions of men to violence. In attempting to destroy each other, they destroyed what was left of the ecology of the island.

In many places throughout the world the first act of this tragedy has already taken place and the final denouement is threatening. In Madagascar, Bali, Indonesia, and also many continental locations, the destruction of forest cover signals impending disaster. It is not just the loss of the trees themselves (although that is a tragedy in itself—the destruction of the very species that sheltered the childhood of man). It is the damage done to the whole symbiotic system of soil, vegetation, rainfall, and sunlight. It is as though a sudden change in ocean level decreased the amount of sunlight reaching a coral reef. Then the zooxanthellae would be wiped out, the coral polyps die, and the fish populations become decimated. So it is when mankind alters drastically the intricately balanced interdependent relationships that have evolved.

Because the Earth is an island isolated in space, over-population cannot be relieved by migration. Although there are science fiction dreams of colonies on Mars or the moon, the possibilities that mankind has available today are much more restricted. There is a tendency to believe that miraculous solutions will be invented by technology—just as fire was used to make more food available to growing populations. But this invention led to the disruption of the elaborately integrated factors that had created a favorable environment for life. We must learn to see each problem as a part of the whole web of life. As John Muir remarked, "When we try to pick out anything by itself, we find it hitched to every-thing else in the universe."

I watched an interview on television with a young woman whose husband was a lumberman in the forests of the American Northwest. One of the great stands of ancient redwood trees was threatened, and the only law that could save it involved the endangered status of the spotted owl. "Is an owl more important than people?" asked the young woman passionately, with absolutely no understanding of the fact that the habitat of the owl must be protected in order to preserve the species and no understanding of the crucial role that forests play in the whole ecology that supports people as well as the spotted owl. Many people, in fact, suffer from this same misunderstanding. Because there is a law that protects an endangered species by preserving its habitat, this technicality is evoked to save an important portion of the natural environment that does not have adequate legal protection in its own right. Unfortunately, it is easier to pass legislation preserving a single species than a magnificent forest, because the economic threat and loss of jobs does not appear to be as significant. But the result is that the objective appears to be so trivial that the cause loses support among the general population. Clearly, the issues should be squarely faced and there should be laws protecting endangered habitats. In so doing, endangered species would be protected as well as the myriad other forms of life that share in the ecosystem.

The method of the physical sciences has traditionally been to simplify a problem to be studied, to isolate the phenomenon in question, and to control all the other variables. However, this method, which has been wonderfully successful in many areas, does not work in studying the relationship between living things and their environment. The factors are so interwoven that one strand cannot be removed without disturbing all the others. If we are ever to preserve an environment that satisfies the needs of the whole web of life on Earth, we must stop thinking in terms of solving isolated problems in a mechanical way and more in terms of understanding the total biotope. This will involve training minds that are capable of evaluating many aspects simultaneously and making judgments based on this broader view.

Perhaps we should treat the whole island Earth as a nature preserve, not in the sense of "do not touch" like the Galápagos, but by a careful consideration of the all the consequences of each interference with nature. With computers we are beginning to have the technology to deal with these complex problems.

The observation of life on islands has given mankind glimpses into the principles of evolution. The survival of the fittest was interpreted to mean that the future belongs to the strongest, the fleetest, the most aggressive, and man pursued this goal with outstanding success. But if we look carefully at the history of life on islands, we see that this is not the final word. There are deeper secrets that have not been explored. The dinosaurs were among the most able predators that the Earth has ever seen. They were the largest, strongest, and best equipped, with lethal weapons like their killer claws. But were they the fittest? We have found their bones in gravel banks beneath the sediment of centuries on the windy shores of Madagascar, while their nearest relatives—delicate and beautiful creatures like the skylark and the hummingbird—have inherited and pass on the precious gift of life. The coral atoll and not the dinosaur has been the ultimate survivor.

Symbiotic relationships are the building blocks on which nature constructs larger and more capable forms of life. We have seen how flowers and trees and fruits made important contributions to the development of mankind. They continue to provide for our well-being in ways that we do not totally understand. It is their presence that has inspired a large part of our literature, music, and art—the sense of beauty and the joy of fulfillment.

The formation of a community must extend beyond our own species to incorporate many other living things. We need the close association with other beings who share with us the mysterious privilege of being alive. We need to increase our understanding of how the many factors work together to create a self- regulating system maintaining the conditions in which life can flourish. Only then can we fulfill our role in this process.

Although many aspects of the planet have been seriously damaged, Gaia is still vibrant, pulsing with life. As Lewis Thomas said, "It is the only exuberant thing in this part of the cosmos. It has the organized, self-contained look of a live creature, full of information, marvelously

skilled in handling the sun." We know that nature is wonderfully resilient. Given proper care, time and nature
will heal the scars. Like the coral atoll, an organism can
recover and rebuild a web of living things. It is not too
late to save the island Earth, but time is growing short.

Notes

Introduction

1   The author of this verse is noted as R. Field but no further information is available.

5   *Volcanic eruptions in the Mediterranean*, as described in Bullard, *Volcanoes of the Earth*, pp. 308–310.

6   *Discovery of guyots*, from Hess, "Drowned Ancient Islands of the Pacific Basin," pp. 772–775.

6   *Volcanoes in or near the sea*, as described by Francis, *Volcanoes*, p. 14.

9   *New insights*, see, for example, MacArthur and Wilson, *The Theory of Island Biogeography*.

Chapter 1   The Hawaiian Archipelago

13   *Poe quote* from "To One in Paradise."

14   *Activity of Mauna Loa*, in Frierson, *The Burning Island*, p. 5.

14   *Age of the Hawaiian chain* as reported in Strahler, *Principles of Physical Geology*, p. 185.

15   *Scientists explore a new seamount*, Broad, *New York Times* (Oct. 8, 1996), p. B7.

16   *Sacrifices to Pelé*, as reported in Bird, *Six Months in the Sandwich Islands*, pp. 76–77, and Frierson, p. 121.

16   *Spiders colonize new lands*, as described in Carson, *The Sea Around Us*, pp. 90–92.

20   *Endemic species of plants in Hawaii*, in Merlin, *Hawaiian Forest Plants*, p. 57.

20   *Endemic flora and fauna of Hawaii*, ibid., pp. 56–57.

23   *Early Hawaiians relationship with the rest of nature*, in Frierson, p. 77.

23   *Bird feathers used as decoration*, as described in Culliney, *Islands in a Far Sea*, pp. 82–83, and Merlin, pp. 56–57.

25   *Ola Pua*, mentioned in Gleasner and Gleasner, *Hawaiian Gardens*, p. 28.

27   *Pollution of Kaneohe Bay*, as reported in Johannes, "Life and Death of the Reef."

Chapter 2   Easter Island

31   *Shakespeare quotation* from *A Midsummer Night's Dream*, Act II, Scene 2.

33   *Translation of Rongo-rongo tablets*, in Fischer, "Preliminary Evidence for Cosmogonic Texts in Rapanui's *Rongorongo* Inscriptions."

33   *Similarity to South America*, as described in Heyerdahl, *Aku Aku*.

34   *Study of bones*, as reported in "New Easter Island Data," *Science Digest* (Oct. 1982), p. 48; and Conniff, "Easter Island Unveiled," p. 77.

34   *Basque crews on Spanish galleons*, in Johanna Pinneo, "Europe's First Family," *National Geographic* (Nov. 1995), pp. 74–97.

35   *Timing of latest volcano*, as stated in Dos Passos, *Easter Island*, p. 125.

36   *NASA measurements*, in Dunn et al., "The Motion of Easter Island from Lageos Laser Ranging."

37   *Cores showing pollen*, as described in Dos Passos, *Easter Island*, p. 130.

40   *Oral history*, as related in Englert, *Island at the Center of the World*, pp. 130–134.

41   *Englert quotations*, ibid., pp. 132–133.

41   *Carbon dating*, ibid., p. 134.

42   *Human skulls and gnawed bones*, mentioned in Coniff, "Easter Island Unveiled," p. 77.

44   *Behren's description*, as quoted in Dos Passos, *Easter Island*, pp. 12–13.

44   *Captain Cook's impression*, ibid., p. 50.

45   *La Perouse's impression*, ibid., pp. 55–57.

45   *Pierre Loti's impression*, ibid., pp. 73–94, 99.

46   *100 acres a minute*, estimate cited in Cleveland, *The Global Commons*.

Chapter 3   The Galápagos Archipelago

49   *Aurelius quotation* from *Meditations*, IV, 36.

51   *The Galápagos Hydrothermal Expedition*, in Corliss and Ballard, "Oases of Life in the Cold Abyss."

54   *Darwin quotation*, in Darwin, *The Voyage of the Beagle* (1958), pp. 325–326.

56   Ibid., pp. 327–328.

Chapter 4   Bali and Lombok

Chapter 5   The Islands of Indonesia

84 *Tsunamis*, as described in Vitaliana, *Legends of the Earth*, pp. 148–151.

84 *Description of Krakatoa eruption*, in Dott and Batten, *Evolution of the Earth*, p. 1

85 *Eruption of Krakatoa*, as described in Francis, *Volcanoes*, pp. 69–81.

85 *Eruption of Tambora*, ibid.

86 *The ring of gold*, ibid., p. 279.

91 *Periwinkle species in Madagascar*, in Jolly, "Madagascar," p. 173.

Chapter 6 Madagascar

93 *Shakespeare quotation* from *Merchant of Venice*, I:1.

94 *Fossils discovered in Madagascar*, in Sampson, Krause, and Forster, "Madagascar's Buried Treasure"; and Goodman and Patterson, eds., *Natural Change and Human Impact in Madagascar*, pp. 3–43.

94 *Similarity to dinosaurs in South America*, in *Bulletin of the Field Museum* (March/April 1997): 1.

95 *Timing of rifting*, in Sampson, Krause, and Forster, "Madagascar's Buried Treasure," pp. 24–27. Sampson et al., "Predatory Dinosaur Remains from Madagascar," posits a revised paleogeographic reconstruction.

96 *Fossils found in Madagascar*, in Sampson, Krause, and Forster, "Madagascar's Buried Treasure," pp. 24–26.

97 *Ancestors of the lemurs*, in Quammen, *The Song of the Dodo*, p. 241.

97 *Number of lemur species*, in Goodman and Patterson, eds., *Natural Change and Human Impact in Madagascar*, pp. 218–220.

98 *Number of endemic species*, in Sampson, Krause, and Forster, "Madagascar's Buried Treasure," p. 24.

98 *Commerson's comment*, as quoted in Jolly, "Madagascar," p. 149.

106 *The elephant bird*, as described in Quammen, *The Song of the Dodo*, p. 17.

107 *Wallace statement*, in Wallace, *The Geographical Distribution of Animals*, pp. 370–371.

108 *Wallace quotation*, from Northrop, Interview with Wallace, *The Outlook*.

109 *Cost of rice*, in Jolly, "Madagascar," p. 177.

146   *Sixteen trillion atoms an hour*, crystal formation
       described in Desautels, *The Mineral Kingdom*, p. 49.
147   *Eiseley quotation*, in Eiseley, *The Immense Journey*,
       p. 27.

Chapter 9   Crete and Santorin

149   *Carson quotation*, in Carson, *The Sea Around Us*, p.
       85.
151   *Origins of the Minoans*, in Judge, "Greece's Brilliant
       Bronze Age," p. 169.
152   *Minoan writing*, in Mellersh, *Minoan Crete*, pp.
       40–41.
154   *No evidence of human sacrifice*, ibid., p. 108.
154   *Three earthquakes per century*, ibid., p. 129.
155   *Explanation of the Minoan decline*, in Marinatos,
       "Thera: Key to the Riddle of Minos."
155   *Ash layer in eastern Mediterranean*, in Dott and
       Batten, *Evolution of the Earth*, p. 1.
155   *Date of eruption*, in Vitaliano, *Legends of the Earth*,
       p. 204. Note that radiocarbon dating has an error
       range of plus or minus 43 years.
156   *Human sacrifices*, in Sakellarakis and Sapouna-
       Sakellarakis, "Drama of Death in a Minoan Temple."
157   *Comparison with Krakatoa*, in Williams, "Crete:
       Cradle of Western Civilization," p. 180.
157   *Four inches of ash*, as estimated in Vitaliano, *Legends
       of the Earth*, p. 191.
159   *Description of tsunamis*, ibid., pp. 148–151.
160   *Egyptian writings of about 1500 B.C.*, in Dott and
       Batten, *Evolution of the Earth*, p. 2.
161   *Bible quotation* from Exodus 9:23–25.
162   *Geologic explanation of the Exodus*, as described in
       Vitaliano, *Legends of the Earth*, pp. 252–271.

Chapter 10   Iceland

165   *Byron quotation*, from *Childe Harold's Pilgrimage*,
       chapter 4, stanza 182.
167   *Krafft quotation*, in Krafft and Krafft, *Volcano*, p. 173.
168   *Volcanic cone on Mars*, in Pollack, "Mars," p. 111.
170   *Eruption on Iceland in 1996*, in Oeland, "Iceland's
       Trial by Fire," pp. 58–73.

171 *Description of the rift beneath Iceland*, Dott and Batten, *Evolution of the Earth*, p. 138.

173 *Viking settlements in Greenland*, in Bryson and Murray, *Climates of Hunger*, pp. 67–71; and Garner and Rosing, "Lost Norse Mystery," pp. 4–9.

174 *Condition of the bodies of the Vikings*, in Schneider and Londer, *The Coevolution of Climate and Life*, p. 112.

174 *Greenland today*, personal communication with Eugene Parker, professor of astronomy and physics, University of Chicago, Sept. 5, 1989.

176 *Cyclical changes in Earth's orbit*, in Hays, Imbrie, and Shackleton, "Variations in the Earth's Orbit," pp. 1083–1087.

177 *Deep-sea vents*, in Sullivan, "Deep-sea Life Is Found Flourishing on Sulfur from Ocean's Volcanoes," p. 33.

178 *Heat and water emerging from the vents*, in Waldrop, "Ocean's Hot Springs Stir Scientific Excitement," pp. 30–33.

Chapter 11 The Bahama Islands

181 *Ovid quotation* from *Metamorphoses*, XV.

182 *Formation of the islands and banks*, in Dott and Batten, *Evolution of the Earth*, p. 253.

183 *Bird species in the Bahamas*, in Bond, *Birds of the West Indies*, pp. 2–3.

184 *History of the Lucayan people*, in Albury, *The Story of the Bahamas*, pp. 13–19.

184 *Quotation from Columbus's journal*, ibid., p. 32.

Chapter 12 Coral Atolls

199 *Shakespeare quotation* from *The Tempest*, I:2.

201 *Dredging from tops of guyots*, in Dott and Batten, *Evolution of the Earth*, p. 141.

203 *Diurnal rings in coral*, ibid., p. 248.

206 *The coral reef as a superorganism*, as described in Chester, *Living Corals*, pp. 17–23.

207 *Depth of volcanic base at Eniwetok*, in Dott and Batten, *Evolution of the Earth*, p. 141.

208 *Evacuation of the inhabitants of Bikini*, in Ellis, "Bikini: A Way of Life Lost," p. 814.

209 *Power of Bravo*, ibid., p. 813.
210 *Atom tests on Bikini*, in Teller and Latter, "Our Nuclear Future."
210 *Life on Kili*, in Ellis, "Bikini: A Way of Life Lost," p. 815.
210 *Fate of the ships at Bikini*, in Eliot, "Nuclear Graveyard."
211 *Decontaminating the topsoil of Bikini*, ibid.
212 *Hydrogen bomb on Eniwetok*, in Menard, *Islands*, p. 206.
212 *Looking like Swiss cheese*, as reported in the *Boston Globe* (May 12, 1980).
213 *Rainbow Warrior II*, as reported in Spielman, "Attack on Rainbow Warrior II," p. 2.
213 *Radioactivity in plankton*, as reported in the *Boston Globe* (Dec. 13, 1990).
217 *Sacks quotation*, from Sacks, *The Island of the Colorblind*, p. 38.
218 *Asthma on Tristan da Cunha*, as reported in Monmaney, "Gene Sleuths Seek Asthma's Secrets on Tristan da Cunha," p. A12.
219 *Captain Cook's visit to Bora Bora*, in Villiers, *Captain James Cook*, p. 139.

Chapter 13   Atlantis
225 *Milne quotation* from the poem "Halfway Down."
227 *Plato quotation* from Timeaus, ii:517.
233 *Continental shelves*, in Marden, "Man's New Frontier," pp. 495–531.
234 *Changes in sea level at end of the last ice age*, in Dott and Batten, *Evolution of the Earth*, p. 429.
235 *Two Creeks and Allerod Retreat*, ibid., pp. 434, 443.
235 *Changes in level of North Sea*, ibid., p. 8.
236 *Dogger Bank*, as described in Carson, *The Sea Around Us*, pp. 72–73.
239 *Discovery of Alexandria's royal quarter*, in Jehl, "Down Among the Sewage: Cleopatra's Storied City."
240 *Recent finds in the harbor*, Public Broadcasting System, *Nova*, "Treasures of the Sunken City."
240 *Earthquake in the fourth century A.D.*, in Jehl, "Down Among the Sewage: Cleopatra's Storied City."

Chapter 14   The Earth

246   *Thomas quotation*, from Thomas, *The Lives of a Cell*, p. 145.

247   *Gaia theory*, in Lovelock, *Gaia*, preface, p. vii.

247   *Composition of the atmosphere*, ibid., chapter 5.

247   *Salt content of the sea*, ibid., p. 81.

248   *Lovelock quote on climate*, ibid., pp. 18, 19.

248   *Gaia theory*, ibid., chapter 3.

250   *Eiseley quotation*, in Eiseley, "How Flowers Changed the World," *The Immense Journey*, p. 77.

258   *Muir quotation*, in Muir, *Gentle Wilderness*, p. 146.

261   *Thomas quotation*, from Thomas, *Lives of a Cell*, p. 145.

References

Albury, Paul. *The Story of the Bahamas*. London: Macmillan Education Ltd., 1975.

Audubon Society. *Field Guide to North American Seashore Creatures*. New York: Alfred A. Knopf, 1981.

———. *Field Guide to North American Fishes, Whales, and Dolphins*. New York: Alfred A. Knopf, 1981.

Bird, Isabella. *Six Months in the Sandwich Islands*, 1875. 7th ed. C. E. Tuttle, 1974.

Blair, Lawrence, and Lorne Blair. *Ring of Fire*. London: Bantam Press, 1988.

Bond, James. *Birds of the West Indies*. Boston: Houghton Mifflin, 1993.

Broad, William J. "Scientists Explore a New Seamount." *New York Times* (October 8, 1996): B7.

Bryson, Reid A., and Thomas J. Murray. *Climates of Hunger*. Madison: University of Wisconsin Press, 1977.

Buffon, Comte de, *Natural History*. London, 1812.

Bullard, Fred M. *Volcanoes of the Earth*. Austin: University of Texas Press, 1976.

Carson, Rachel, *The Sea Around Us*. New York: Oxford University Press, 1951.

Chernush, Akosh. "Dazzling Jewels from Muddy Pits Enrich Sri Lanka." *Smithsonian* 11 (June 1980): 68–74.

Chester, Richard. *Living Corals*. New York: Clarkson N. Potter, 1979.

Cleveland, Harlan. *The Global Commons: A Policy for the Planet*. Aspen Institute, 1990.

Conniff, Richard. "Easter Island Unveiled." *National Geographic* (March 1993): 54–79.

Corliss, John B., and Roger D. Ballard. "Oases of Life in the Cold Abyss." *National Geographic* (October 1977): 441–453.

Cottrell, Leonard. *Lost Cities*. New York: Rinehart & Co., Inc., 1957.

Culliney, John L. *Islands in a Far Sea: Nature and Man in Hawaii*. San Francisco: Sierra Club Books.

Darwin, Charles. *Journal of Researches into the Geology and Natural History of the Various Countries Visited by H.M.S. Beagle, 1832–1836*. London, 1839. Reprinted as

The Voyage of the Beagle, complete and unabridged edition. New York: Bantam Books, 1906.

———. *On the Origin of the Species by Means of Natural Selection*. London: J. Murray, 1st ed. 1859; 6th ed., 1872.

Darwin, Erasmus. *Zoonomia, or the Laws of Organic Life*. London, 1794.

Desautels, Paul E. *The Mineral Kingdom*. New York: Madison Square Press, 1968.

Dingus, Lowell, and Timothy Rowe, *The Mistaken Extinction*. New York: W. H. Freeman, 1997.

Donnelly, Ignatius. *Atlantis: The Antediluvian World*. New York: Harper & Brothers, 1882. Reprint Dover Publications, Inc. 1976.

Dos Passos, John. *Easter Island: Island of Enigmas*. Garden City, N.Y.: Doubleday & Co., Inc., 1971.

Dott, H. Robert, Jr., and Roger Batten. *Evolution of the Earth*. New York: McGraw-Hill, Inc., 1971, 1976.

Dunn, P. J., et al. "The Motion of Easter Island from Lageos Laser Ranging," Greenbelt, Md.: NASA Goddard Space Flight Center.

Eiseley, Loren. *Darwin's Century*. Garden City, N.Y.: Doubleday, 1958.

———. *The Immense Journey*. New York: Random House, 1957.

Eliot, John L. "Nuclear Graveyard." *National Geographic* (June 1992): 70–82.

Ellis, William S. "Bikini: A Way of Life Lost." *National Geographic* (June 1986): 813.

Englert, Father Sebastian. *Island at the Center of the World*. Translated and edited by William Mulloy. New York: Charles Scribner's Sons, 1970.

Fischer, Stephen Roger. "Preliminary Evidence for Cosmogonic Texts in Rapanui's Rongorongo Inscriptions." *Journal of the Polynesian Society* 104 (1995): 313–321.

Francis, Peter. *Volcanoes*. London: Penguin Books Ltd., 1976.

Frierson, Pamela. *The Burning Island*. San Francisco: Sierra Club Books, 1991.

Garner, Fradley, and Jens Rosing. "Lost Norse Mystery." *Oceans* 10 (March 1977): 4–9.

Gillispie, Charles C., "The Formation of Lamarck's

Evolutionary Theory," *Archives Internationales d'histoire des sciences* 9 (1956): 323–338.

Gleasner, Bill, and Diana Gleasner. *Hawaiian Gardens.* Honolulu: Oriental Publishing Company, 1977.

Goodman, Steven M., and Bruce Patterson, eds. *Natural Change and Human Impact in Madagascar.* Washington, D.C.: Smithsonian Institution Press, 1997.

Gordon, Michael R., "Reversal of Forces in Gulf: Iraqis Now Set for Defense," *New York Times* (Nov. 4, 1990): 1.

Graedel, Thomas E., and Paul J. Crutzen. *Atmosphere, Climate, and Change.* New York: Scientific American Library, 1995.

Green, Timothy S. "Diamond Diggers in Nambia." *Smithsonian* 12 (May 1981): 49–57.

Hayes, J. D., John Imbrie, and N. J. Shackleton. "Variations in the Earth's Orbit: Pacemaker of the Ice Ages." *Science* 194 (Dec. 10, 1976): 1083–1087.

Hess, H. H. "Drowned Ancient Islands of the Pacific Basin." *American Journal of Science* 244 (November 1946): 772–791.

Heyerdahl, Thor. *Aku Aku.* Chicago: Rand McNally, 1958.

Huxley, Julian. *Evolution in Action.* New York: Harper & Brothers, 1953.

———. *Evolution: The Modern Synthesis.* New York: John Wiley & Sons, Inc., 1964.

Jehl, Douglas. "Down Among the Sewage: Cleopatra's Storied City," *New York Times* (October 29, 1997).

Johannes, R. E. "Life and Death of the Reef." *Audubon* 78 (September 1976): 38–55.

Jolly, Alison. "Madagascar: A World Apart." *National Geographic* (February 1997): 148–183.

Jones, Stuart E. "Spices: The Essence of Geography." *National Geographic* (March 1949): 401–420.

Judge, Joseph. "Greece's Brilliant Bronze Age." *National Geographic* (February 1978): 142–185.

Krafft, Maurice, and Katia Krafft. *Volcano.* New York: Harry N. Abrams, 1973.

Lovelock, James. *Gaia: A New Look at Life on Earth.* New York: Oxford University Press, 1979.

Malthus, Thomas, *An Essay on the Principle of Population as It Affects the Future Improvement of Society.* London, 1798.

MacArthur, Roger H., and Edward O. Wilson. *The Theory of Island Biogeography*. Princeton, N.J.: Princeton University Press, 1967.

McCarry, John. "Island of Quiet Success." *National Geographic* (April 1993): 111–132.

McCurry, Steve. "Sri Lanka." *National Geographic* (January 1997).

Marden, Luis. "Man's New Frontier." *National Geographic* 153, no. 4 (April 1978): 495–531.

Marinatos, Spyridon. "Thera: Key to the Riddle of Minos." *National Geographic* (May 1972): 170–180.

Mellersh, H. E. L. *Minoan Crete*. New York: G. P. Putnam's Sons, 1967.

Menard, H. W. *Islands*. New York: Scientific American Library, 1986.

Merlin, Mark David. *Hawaiian Forest Plants*. Honolulu: Oriental Publishing Company, 1976.

Monmaney, Terence. "Gene Sleuths Seek Asthma's Secrets on Tristan da Cunha." *Los Angeles Times* (April 30, 1997): A12.

Moorehead, Alan. *Darwin and the Beagle*. New York: Harper & Row, 1969.

Muir, John. *Gentle Wilderness: The Sierra Nevada*. Edited by David Brower. New York: Ballantine Books, 1967.

Northrop, W. B., Interview with Wallace. *The Outlook* 22 (November 1913).

Oeland, Glenn. "Iceland's Trial by Fire." *National Geographic* (May 1997): 58–73.

Pollack, James B. "Mars." *Scientific American* 233 (September 1975): 111.

Public Broadcasting System. "Treasures of the Sunken City: The Seventh Wonder of the Ancient World," *Nova* episode 2417 (June 8, 1998).

Quammen, David. *The Song of the Dodo*. New York: Charles Scribner's Sons, 1996.

Rosenzweig, Michael. "Tempo and Mode of Speciation," *Science* 277 (September 12, 1997): 1622.

Sacks, Oliver. *The Island of the Colorblind*. New York: Alfred A. Knopf, 1997.

Sakellarakis, Yannes, and Efi Sapouna-Sakellarakis. "Drama of Death in a Minoan Temple." *National Geographic* (February 1981): 205–222.

Sampson, Scott D., David W. Krause, and Catherine A. Forster. "Madagascar's Buried Treasure." *Journal of Natural History* (March 1997): 24–27.

Sampson, Scott D., et al. "Predatory Dinosaur Remains from Madagascar: Implications for the Cretaceous Biography of Gondwana." *Science* (May 15, 1998): 1048–1052.

Schneider, Stephen, and Randy Londer. *The Coevolution of Climate and Life*. San Francisco: Sierra Club Books, 1984.

Scidmore, Eliza. "Adam's Second Eden." *National Geographic* (February 1912): 105–173.

Smil, Vaclav. *Cycles of Life: Civilization and the Biosphere*. New York: Scientific American Library, 1997.

Spielman, Peter James. "Attack on Rainbow Warrior II." *Boston Globe* (May 12, 1995).

Strahler, Arthur N. *Principles of Physical Geology*. New York: Harper & Row, 1977.

Sullivan, Walter. "Deep-sea Life Is Found Flourishing on Sulfur from Ocean's Volcanoes." *New York Times* (April 25, 1982): 33.

Teilhard de Chardin, Pierre. *Le phénomène humain* (1955). Translated as *The Phenomenon of Man* by Bernard Wall. New York: Harper & Brothers, 1959.

Teller, Edward, and Albert Latter. "Our Nuclear Future: Facts, Dangers and Opportunities." *Life* magazine (February 10, 1958).

Thomas, Lewis. *The Lives of a Cell*. New York: Viking Press, 1979.

Villier, Alan. *Captain James Cook*. New York: Charles Scribner's Sons, 1967.

Vitaliano, Dorothy B. *Legends of the Earth*. Bloomington: Indiana University Press, 1973.

Waldrop, Mitch. "Ocean's Hot Springs Stir Scientific Excitement." *Chemical and Engineering News* (March 10, 1980): 30–33.

Wallace, Alfred Russel. *The Malay Archipelago: The Land of the Orang-utan, and the Bird of Paradise. A Narrative of Travel with Studies of Man and Nature*. New York: Harper & Brothers, 1869.

———. *The Geographical Distribution of Animals*. Vol. II. London and New York, 1876.

White, Alan, and Bruce Epler. *Galapagos Guide.* Academy Bay: Charles Darwin Foundation, 1972.

Williams, Maynard Owen. "Crete, Cradle of Western Civilization." *National Geographic* (February 1978).

Wilson, Edward O. *Biophilia.* Cambridge, Mass.: Harvard University Press, 1984.

Yoon, Carol Kaesuk. "Hapless Iguanas Float Away and Make Biological History." *New York Times* (Oct. 8, 1998).

Young, Louise B. *The Blue Planet.* Boston: Little, Brown, and Company, 1983.

Young, Louise B., ed. *Evolution of Man.* New York: Oxford University Press, 1970.

———. *The Mystery of Matter.* New York: Oxford University Press, 1965.

Index